TETHERED

TETHERED

Technology, Faith & the Illusion of Self-Sufficiency

C.M. COLLINS

Walnut Creek, California

Library of Congress Cataloging-in-Publication Data available on request.

ISBN-13 978-0-9897389-8-9
ISBN-10 0989738981

Wind in the Reeds Publishing
Walnut Creek, CA
www.windinthereedspublishing.com
www.windinthereedspub@gmail.com

Cover photograph and design by Ren Reed.

Printed in the United States of America.

10 9 8 7 6 5 4 3 2 1

To Martha

TABLE OF CONTENTS

PART III. EXPRESSION

INTRODUCTION

THE FLOWER POWER GENERATION

I pretty much missed the sixties—even though I was there.

Now this is an old joke, usually having something to do with a person's memories of the era being addled by their ingestion of illegal substances. I missed the sixties because I was just a little kid, too young to understand the changes going on in the world outside our home. We lived in Santa Clara, California, just a stone's throw from San Francisco and Berkeley, but my family's ties to the counterculture were thin at best. They came primarily through visits from my older cousin Michael.

When Michael was seventeen, he had tuned in, turned on, and run away from home. He had grown his hair long and started living a hippie lifestyle, selling cheap jewelry made out of colorful beads. Nobody knew exactly where he lived or where he was at any given time until he'd show up completely out of the blue. I can still remember playing out on the front lawn one day with my sister Melinda and looking up to see a scruffy looking teenager coming around the corner down the street. As he got closer, we recognized him. "It's Mike!" we yelled, and ran to hug him.

He stayed for a few days, hanging out with the family. He showed my sister how to make beaded earrings, and helped me build my Legos all the way to the ceiling of my bedroom. We laughed and laughed when they all came crashing to the floor. We woke up one morning to find him gone again, off to wherever it was that hippies did whatever it was they did. We loved him.

There *were* a few other signs of countercultural rebellion around our house. My sister had a brightly colored *Flower Power* beach towel my mom bought her at JCPenney, and my parents went to see *Hair* and brought home the soundtrack. At our Baptist church, the pastor's son ran off to Canada to avoid the draft. But that was really about it. When my dad got a teaching job at the University of Costa Rica in the early seventies, we left the States for several years. By the time we got back, disco was dawning and the Age of Aquarius was over.

Looking back on it now, the Bay Area in the late 1960s and early 1970s was a unique place to be: the hippies, the Free Speech Movement, Kesey's Merry Pranksters, the Black Panthers, Vietnam, LSD, the Jesus Freaks, the *Whole Earth Catalog*. It was either the end of civilization as people knew it, or the dawn of a new era of consciousness that would change the world forever.

It turns out it was neither. Or both.

What strikes me most about the era was its earnestness. It seems hard to imagine now, but many of the people performing the LSD experiments at Stanford really thought they might bring an end to war by cracking open the doors of perception. The Jesus Freaks handing out tracts in Haight-Ashbury were genuinely concerned for your eternal destiny. Everyone everywhere was trying to make the world a better place for children and all other living things.

So, did it work? Did the counterculture alter the course of history forever? Let's look at the world today:

- The war in Vietnam is now the war in Afghanistan.
- The Civil Rights protests in the streets have turned into still more Civil Rights protests in the streets.
- The drug experimentation of the era has mutated into the meth and heroin epidemics of today.
- 1960s protest songs are available for download on iTunes, courtesy of Apple, one of the world's largest corporations.
- The free speech radicals on college campuses have become today's rigid moralists, demanding spaces safe from all conflicting thought.
- The Jesus Movement faded, and the Holy Spirit was channeled into right wing politics.
- And most importantly for our discussion, the personal computer, which arose out of the countercultural ethos with promises of freeing the individual from the tyranny of the System, has morphed into today's cell phone and internet addictions.

On the positive side, the counterculture did raise some very important existential questions, ones that continue to resonate deeply through society to this day, much to our benefit. Things *are* better now for many. It's just that the overarching transformational promise of the era was not achieved, despite all the genuine effort.

Why not? Why didn't it work? I don't think the counterculture itself was to blame. For the most part, it was made up of kids—kids who were playing the time-honored role of pointing out the hypocrisy of their elders. They were honestly trying. But none of those students who chained themselves together in the streets to protest governmental corruption could have envisioned the NSA's ability to track and record our every move and communication in real time. And if you were to put one of them in a time machine and send them to a college campus today, they'd experience total culture shock. *Why is everyone walking around with their heads bowed, looking down at little boxes? Why do two friends sit next to each other and not talk? Where is the community?* This was not what they wanted. Nor did they cause it.

The trouble was they were fighting the wrong enemy. They could not see that the desires that led them to seek the overthrow of the System were the same ones that led to the creation of the System in the first place. They believed they were fighting against war, political corruption, religious hypocrisy, racism, sexism and a lack of freedom, but did not understand that these were merely the outward manifestations of a deeper human and spiritual dilemma. Their solutions, therefore, ended up being palliative, rather than curative.

This same illusory mindset is evident today, primarily in the world of technology. We are constantly being bombarded with grandiose promises of the glorious technological utopia that is right around the corner. No one knows exactly what it will look like, the picture is still vague, but it's going to be *fantastic*, we are assured. All our problems will finally be solved. It will be the dawn of a new era of consciousness that will change the world forever . . . uh. . . and just where have we heard this all before? Just like in the old parable, the emperor is still not wearing any clothes.

Since our basic problem is not technological, its solution cannot come from technology.

Every effort is being made to obscure this fundamental

truth, however, because we desperately want to believe that *we* will be the generation that finally gets it right. We forget that for every technological advance we make, a whole host of new problems is always created, and that as our powers increase, so do the potential ramifications of our actions. We end up further and further from the things that are truly good for us.

This dilemma is not a new one. In fact, it is very old. We can find it described in many of our ancient myths and stories. We can read about it in more recent works by Aldous Huxley, George Orwell, Jacques Ellul, C.S. Lewis and Neil Postman. All of these prophetic voices are becoming increasingly difficult to hear through the technological chatter, however, primarily because we absolutely love and adore and have placed our faith in this silicon fairy tale where everyone lives happily ever after.

For people of faith and spirituality, therefore, the time has come to ask a series of very important questions: Is there a technological line that you will not cross? Is there a technological line that you *should* not cross? If so, where is it? Why? Have you crossed it already? How will you know? How will you even begin to think about it?

It's important that we ask ourselves these questions before the line actually manifests itself, if it hasn't already. As the saying goes, it's pretty difficult to establish your sexual boundaries when you're already in the backseat of your car on a Friday night. If we are not ready when the time comes, we may miss the moment, and I'm afraid that our lack of foresight will make us complicit in the damage done.

These are the central questions of this book. I do not believe that the answers are very difficult. In fact, they are things that even a child can understand quite easily. It is in their very simplicity that we can be set free. And when we have been set free, we will come to find that it wasn't technology that was the problem in the first place. It was us.

So whatever happened to my cousin Michael? Did he ever find what he was looking for? I'm not sure. For a time, he quit the counterculture, got married and started a career. This proved short lived. The last time we heard from him, he was running away again, this time trying to find happiness somewhere in Spain. Like so many of his generation, indeed of every generation, he seemed to be on a perpetual quest for a meaning that lay just beyond his grasp. A bumper sticker says, *All who wander may not be lost.* I hope Michael is not wandering forever.

PART I: THEORY

"God made us plain and simple,
but we have made ourselves very complicated."

Ecclesiastes 7:29

THE THINGS THAT ARE GOOD FOR US

To begin our discussion, we're going to start with one of humanity's oldest stories: *The Garden of Eden.* If you are not used to this sort of thing, don't panic. You will find the story to be very insightful about human nature, whether you happen to believe in its literal truth or not. Besides, we will be reading symbolically, making the truths of the story available to everyone. If you have not read it recently, this would be a good time to do so, so that it is fresh in your mind for our discussion.

The Garden of Eden is a story of lost potential. It tells us who we were created to be and who we have actually become, and why the two are so different. Interestingly, it also has a lot to say about our relationship to technology. Most of all, for our purposes, it shows us the things that are truly good for us, and why we don't have them.

There are actually two different stories at the beginning of Genesis. The first story, which is told in poetic form, describes the six days of creation (Genesis 1). In it, we learn that both the male and the female are made equally in the image of God and are responsible for the care of the world. When God is finished creating, he looks at everything he has made and is well pleased. He rests on the seventh day.

The Garden of Eden (Genesis 2-3) is much more interesting, as a story. In it, the man is formed out of the soil of the ground. God breathes into his nostrils and he begins to live. The symbolism is deceptively simple: we are made of soil—ashes to ashes and dust to dust—but we also contain an element of the Spirit.

This means that while we are biological creatures, we also have a spiritual side that is not biological in origin. This directly opposes the idea that human consciousness is solely a matter of brain functioning. If the story is true, then all of our modern day attempts to prove the "brain-only" hypothesis will inevitably fall short. Our scientific instruments, as powerful as they are, are simply the wrong tools for the job. How does one measure God? It also means that we will never be able to fully replicate a human being. What machine will ever be able to perform the *breathing* part? Of course at this point, we can't even define what consciousness *is,* much less simulate it. Might it be possible that the ancients knew far more than we ever like to give them credit for?

Certainly they had a lot to say about human nature and the nature of God. The story makes several humorous, yet perceptive, observations:

The first of these comes when God gives the man a job to do. Or, we should say, a "non-job". In Eden, the trees grow and produce fruit all by themselves. God tells the man that he is supposed to cultivate the garden, even though there is no work for him to do. It's like the college basketball player in the movie *One on One* who gets paid to watch the gardeners turn the sprinklers on and off—all he has to do is show up in order to keep his scholarship. It's a funny moment in the film and the audience laughs, but in *The Garden of Eden,* the point is more significant. This "non-job" is the first instance of a major theme that runs through the rest of the Bible: God is the one who provides for us; He does not need our efforts. As the man walks through the garden, therefore, and sees it growing magnificently around him, he is constantly reminded of the care God has for him and knows that he has done absolutely nothing to sustain it.

Relationships are key to the entire story. The man is made first, for example, but it quickly becomes apparent that he needs help. *It is not good for man to live alone*, God says, *he needs a suitable companion to help him.* The Hebrew word for "companion" is *ezer*, which conveys the idea of an ideal partner. The word for "suitable" is *kenegdo*, which literally means *according to the opposite of him.*[1] When he sees the woman for the first time, the man is clearly overwhelmed. *Bone of my bone, flesh of my flesh*, he says. The two are a perfect match for each other.

Not only are the man and woman perfectly compatible, but they also live in communion with nature. At one point God brings the animals to the man *to see what he would name them.* We name the things we care for, and are supposed to care for, just as we name our children.

The relationship imagery continues. The first thing God tells the man and woman to do—the first commandment, if you will—is to be fruitful and multiply. This means, strictly by logical inference and deduction from biological necessity, of course, that God is telling them to have sex—a lot. The whole garden environment, then, from the plants to the animals to the people, is very fertile.[2]

And what about the other commandments? This is the Bible, after all. Aren't there lots of sins they need to avoid?

---

[1] Walton, John.

[2] The emphasis in Genesis 1-3 is on relationship mandates, rather than cultural ones. See p. 177.

Aren't there lots of rules? The answer is "no". In the Garden, there is only one rule—they are not to eat the fruit from the tree that gives the knowledge of what is good and what is evil. They can do whatever else they like, including eating from the tree whose fruit gives eternal life.

The story, then, shows us who we are. In the Garden, the man and woman are made of both physical and spiritual elements. They are rational beings who can think. They inherently have language and can speak. They can talk directly with God, and they live in harmony with nature. They are sexual beings who have been created especially for each other. Work is strictly voluntary, since God is the provider and sustainer, and there is only one rule to break, so they can essentially do whatever else they want. We can see why the word *Edenic* means *in a state of perfect bliss and happiness.*

According to the story, we were meant to live:

- In perfect communion with nature,
- In perfect communion with each other,
- In perfect communion with God, and
- In perfect freedom.

You do not need to be religious to see the deep truth here. Our lives thrive best when we live harmoniously with nature, have deep connections with each other, are spiritually in tune, and when we are free. These are the things that are truly good for us.

It is also painfully obvious that this is not the way that humanity generally lives. Instead, we find the exact opposite to be true: we are alienated from nature, constantly at war with each other, and are spiritually adrift much of the time.

Historically and personally speaking, our freedoms have far too often been fleeting. Why aren't we living the lives we were created to live? That comes in the next part of the story.

It all starts with the famous encounter between the woman and the snake. The snake asks, *Did God really say, 'You must not eat from any tree in the garden'?* The woman answers that they are *not to even touch it or they will die.* Notice the subtle addition she makes. God never says they can't touch the fruit—they just aren't supposed to eat it. In her answer, the woman elucidates a fundamental human characteristic: we *love* rules. Where none exist, we invent them. If we don't think the existing ones are strong enough, we add to them. We somehow believe that if we can surround ourselves with enough rules and regulations they will form a shield around us that will protect us from all evil.

Her addendum probably seems like a good idea at first. After all, if you never touch the fruit, you'll certainly never eat it. Because of the addition, however, whoever then touches the fruit will be in the curious position of breaking a rule *without having done anything wrong.* When no punishment is then forthcoming for merely touching the fruit (since it wasn't wrong in the first place), the person will then be all the more likely to go ahead and take a bite out of it too, all rules seemingly being the same. In this way, the creation of extra rules actually leads people further into temptation.

Eventually, rules expand into full-blown moral systems. If not touching the fruit is a good idea, why not add a rule that says to not even look at it? And it sure would be easier to not look at it if there was a fence around the tree. And naturally the fence ought to be the proper height, made out of the proper materials, and painted the proper color. The possibilities are endless. In this way and often despite our best intentions, one simple rule can grow into a massive web of laws that robs us of the freedoms we were created for.

Another by-product of unnecessary rulemaking is excessive pride. The more rules there are, the more things there are for the rule makers to be judgmental about. This has always been a problem among people who are religious. Rules and laws are necessary for the proper functioning of society, of course, but we can end up placing unmanageable burdens on people when we try to force them to conform to our own overly inflated moral agendas. Besides, all of this moralizing does nothing to cure our basic problem. We can't even keep a rule when there is only one of them to break.

And can we please stop blaming the woman for all of this? After all, the prohibition concerning the fruit was originally given to the man, before she was even created. Evidently, a lack of communication between husbands and wives is a very old issue.

The conversation continues. The snake tells her, *You will not surely die, for God knows that when you eat the fruit, your eyes will be opened, and you will be like God, knowing good and evil.* This is a brilliant answer—because it is true. When the man and woman do eventually eat the fruit, they don't immediately die, and their eyes are indeed opened.

The snake's proposition leads her right up to the very edge of the line she should not cross: *You will be like God.* It's the oldest temptation in the book. Now, the woman should just answer, "I already *am* like God. See? Look at me. I was made in his image. I am rational, creative, and social. I am living here in paradise with the animals and my semi-communicative soul mate. I have everything I need. I am free. Go away."

But she does not. Instead, she contemplates the offer.

As she does, she realizes that there is one obvious way in which she is not like God—she does not have the knowledge of good and evil. The story continues, *When the woman saw that the fruit was good for food and pleasing to the eye, and also desirable for gaining wisdom, she took some and ate it.* In doing so, she takes upon herself the power of self-determination. From this point on, she will be the arbiter of what is good and what is evil.

Tragically, however, she has not seen the entire picture. The snake does tell her the truth, but that truth is mixed with a lie. While she has gained a measure of self-sufficiency, she still lacks the power to determine her own destiny. The snake never tells her that to be *fully* Godlike, you have to also eat from the second tree, the Tree of Life.

This is humanity's central predicament. We perceive of ourselves as being Godlike—and we are. But being *like* God is not the same thing as *being* God. The great wisdom of the story is that there are two separate trees. For all of our Godlike greatness that comes from eating from the Tree of Knowledge, we haven't eaten from the Tree of Life, and we *will* die, just as God had warned. Virtually all human endeavor is an attempt to grapple with this reality.

We instinctively know what is good for us: communion with nature, communion with each other, communion with God, and living in freedom. While we have been given all we need to live our lives this way, we are still not satisfied. *You will be like God,* says the snake. We cannot resist. When we eat the fruit and find out that we have been duped, we are left scrambling. "Since we've chosen to be self-sufficient," we say, "we'd better *do* something. It's all up to us now." So we create religions to try to re-establish the connection to God that has been broken. We create political and moral systems to try to mend our broken relationships with each other. We create machines to try to help offset the loss of communion with nature. We even try to change ourselves. It doesn't work. The fruit from the Tree of Knowledge just ends up making us sick. *Knowledge is power*, we like to say, conveniently forgetting that it is also true that *power corrupts.*

Worst of all, the steps we take to help us compensate for our brokenness end up enslaving us even further. Being like God is supposed to set us free. Ironically, in trying to be self-sufficient, we give our freedoms away—right up to the very day that we reach our inevitable destination, death, which is still out of our control. For this, we have no answer.

We have eaten from the Tree of Knowledge, but not from the Tree of Life.

We cannot end our discussion of this section of *The Garden of Eden* without noting a couple of other humorous observations the story makes. After the woman eats the fruit, it says, *She also gave some to her husband, who was with her, and he ate it.* Had he been there the whole time, listening? The story doesn't say. If so, he certainly didn't add much to the conversation. In any case, there is no debate, no deliberation, no discussion for him. He just does what she tells him to do. This might lead us to ask the proverbial question about who wears the pants in the family, except, of course, that neither of them is wearing anything at all. Yet. That comes next: *Then the eyes of both of them were opened, and they realized they were naked; so they sewed fig leaves together and made coverings for themselves.* In one fell swoop, then, we see the tragic, devastating loss of humanity's innocence *and* the start of the fashion industry. There's got to be a life lesson in there somewhere.

Like small children who have done something wrong, the man and the woman hide in shame. Like any parent, God knows what they have done and wants them to own up to it. *Have you eaten from the tree that I commanded you not to eat from?* he asks. The man immediately plays the blame game: *The woman you put here with me—she gave me some fruit from the tree, and I ate it.* Having done something wrong, the man first blames God, then blames the woman. It's a tale as old as time. When the woman is asked to account for what she has done, she too deflects responsibility. *The serpent deceived me, and I ate it,* she says.

A personal anecdote might help get us to the truth that is being illustrated here. One of the common characteristics of people who consider themselves to be visionaries is what I like to call a cult-like *sheen.* The techie gurus are no different. Whenever they are asked about their work, their faces immediately start to change—they get a glassy far-away look and they smile a beatific smile. They stay disturbingly calm even as their words become more and more enthusiastic. If you're not ready for this, it can be unnerving—you're not sure where you're supposed to look. It's always the same: the far away eyes, the trance-like demeanor, the creepy positivity. As they talk it is clear that you too are supposed to get swept up in their feverish visions.

I got a chance to talk with one of these gurus once, and he began to tell me all about the new software he was developing and how it was going to be used to transform the world of interpersonal communications as we knew it, etc. His eyes went glassy. He smiled as he spoke. He began to get the *sheen.* After several minutes of this, I began to worry that he might actually start levitating, so I tossed in a couple of questions. "Does it ever bother you," I asked, "that when people communicate via machines they often lose their ability to talk to one another face to face? Do you ever worry that we might actually be making things *worse?*" He stopped abruptly and was silent for what seemed like a full thirty seconds as the glow slowly drained from his face. I had broken the reverie. I had interrupted his flow. He looked at me blankly, and then quickly excused himself. Like any other cult leader, it was clear that he was not accustomed to having his piety challenged.

The whole encounter was strange. I had just asked a couple of simple questions. But questions lead to answers, and answers to accountability, and none of us really likes that. Not that he owed me any explanation. I was just a nobody from Nowheresville to him. The conversation did lead me to think, however, about how often we are just like the man and the woman in the garden when it comes to technology. The techie gurus come along and offer us their latest fruits, and we almost always just accept them. Part of this is because we have been conditioned to do so (as we will see in a later chapter), and part of it is because we find that their fruit, like that in Eden, is *pleasing to the eye.* But shouldn't we be asking questions and challenging assumptions, especially when the consequences are potentially so dire? We know that the gurus, like the snake, will only give us one side of the story. Shouldn't we at least *doubt?* Does our passivity make us accountable too?

After they have eaten the fruit, God allows the man and the woman to inhabit the world they have chosen. You'd be hard-pressed to find a more accurate description of the state of the world today than the one found in the story. The language here is descriptive, rather than proscriptive:

Instead of communion with nature, they are now alienated from it. The woman and her offspring become enemies of the snake. The ground itself is the enemy of the man. Perfect harmony with the earth is replaced by permanent conflict and difficulty.

Instead of communion with each other, there is now a hierarchy. *Your desire shall be for your husband, and he will rule over you,* it says. In seeking to be wise, they have undone the equality they were granted at creation, and broken the bond that God intended for them.

Instead of having communion with God, they are banished from the Garden forever. They will no longer walk with him and talk with him in such a personal manner. They will continue to communicate, but the intimacy and familiarity of their conversations will forever be changed.

And what about freedom? Work is now no longer optional, but is necessary for survival: *By the sweat of your brow you will eat your food until you return to the ground, for dust you are, and to dust you will return.* Chilling words, but true. Their lives may now be their own, but they will be filled with difficulty and suffering, only to eventually end in death.

The final symbol in the story is an important one. An angel with a flaming sword is placed at the east side of the garden to keep them from ever returning to eat from the Tree of Life. One of the most powerful, universal longings of humanity is a symbolic desire to return to Eden and a state of perfect innocence. This image, however, makes it clear that we can never go back.

The stage is now set for our discussion of technology. *The Garden of Eden* may be religious in origin, but its truths are applicable in far more wide-ranging areas. The story shows us the things that are good for us. It also shows how our own choices keep us from having them, all while we delude ourselves into thinking none of this is our fault. Having started down the path of self-sufficiency, we can never stop, or we won't survive in the world. Little do we understand, however, that all our efforts eventually slip out of our control and enslave us in the end. My term for our dilemma is the *Technological Continuum*, and is the subject of the next chapter.

I HAVE TWEETED NO TWEETS

I have tweeted no tweets
And texted no texts.
I have snapped no chats
And insta'd no grams.
I have flick'd no flickrs
And tumbled no tumblrs.
I've i'd no pads,
Faced no books,
And I have not linked in.
I have absolutely no interest—
In something called pinterest.

But I did notice the beauty
Of the sunrise this morning.

C.M. Collins

THE TECHNOLOGICAL CONTINUUM

It often seems that something noble about the world has been lost. Life can be ruthless, and day-to-day survival is difficult. All the while, the whispers from Eden continue to haunt us: *paradise was real, and it was once within our grasp*. We know we can't go back in time, so we take our yearnings for our idealized past and project them forward into the theoretical future, longing for the coming day when we will finally be able to create a new Eden of our own.

The common term for this is *utopianism*—the desire to create a perfect world on earth. If Eden represents the perfect place for humanity as created by God, then Utopia represents the perfect place for humanity as created by humans.

We live our lives on a continuum between the two extremes:

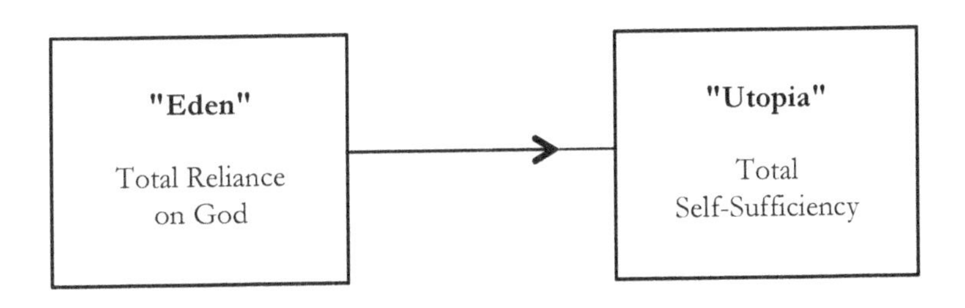

Human history shows that we are on a continual movement to the right. (That's why the arrow is pointing in that direction). Although there have been occasional times when the arrow has slowed or even stopped, these have always been temporary and limited. The overall historical trend has been towards less reliance on God and more dependence on ourselves. Technology has always played a key role in that movement.

This helps us to establish a working definition for technology. Because it is continually changing, it has been hard to capture its essence with a static definition, like one you would find in a dictionary. The idea of the continuum enables us to see that the best way to explain technology is in terms of *movement* and *relationship*.

Using a static definition alone would show us that technology is one of the most important means we have to help us survive in this cruel, cursed world. But that is only part of the story. Technology is also a movement. Every technological "advance" that we make means a move down the line towards greater self-sufficiency. This, then, has implications for relationships. Every movement towards greater self-sufficiency is, by definition, a move away from reliance on God.

This means that technology always has a spiritual dimension.

It may seem counterintuitive to speak of technology in this way, but we have always done so. People have long been aware of the spiritual implications of their tools. Consider the plow. Even as it helped ancient farmers become more efficient, they still understood that they also needed rain, fertile soil and protection from invading insects to help their crops grow. Farming itself was a spiritual act. Their plows and other tools were included in their sacred rituals in hopes that God or the gods would bless them. Because we in the modern world perceive of a plow as being just a hunk of metal, we diminish the ancients by labeling their beliefs as superstitious.

But we are no less superstitious today. We still live in awe of our machines. At the very highest levels, technology today is pursuing the creation of life, the creation of human beings, and ultimately, the elimination of death. In doing so, it is attempting to perform all of the traditional "God functions". Is it any wonder that, as we shall see, so much of modern technology comes wrapped in religious garb?

Technology is where people today now place their supreme faith. Any problems that we have will eventually be solved by the omnipotent technologists and their wonderful machines, we believe. Even something as seemingly ordinary as a smartphone is surrounded by a religious aura. It's a source of wonder (*it works in mysterious ways*), omnipotence (*its power is beyond comprehension*), omniscience (*it is the source of all information*), and even abiding love (*it walks with me and it talks with me*). To us, it is nothing short of miraculous. And walk into virtually any home in the modern world, and you will see the various shrines people have set up to bow and pray to the technological gods they have purchased.

Every technological move we make, therefore, ought to be understood as a spiritual, as well as a temporal act. As we move toward greater and greater self-sufficiency, we move away from dependence on God, and the spiritual dangers increase.

Let's add to the picture:

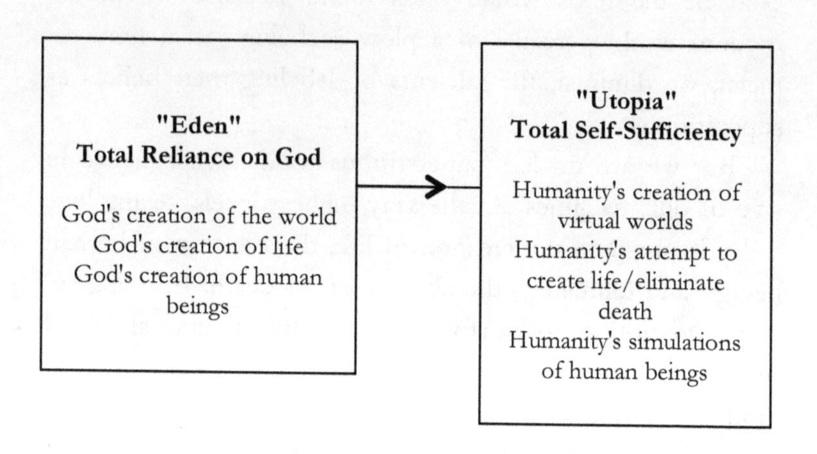

Ironically, all of the talk about how powerful tech has become masks the fact that its agenda is largely motivated by fear. Living east of Eden, we have to constantly worry about our survival. It's a dog eat dog world out there, we say. Only a continual move towards greater power will keep us safe. Might makes right. We can't opt out. The wheels must be kept turning, or we will die. *Fear.*

Economically, we're told that if we don't keep updating our machines and skills, we'll be left behind. We won't be able to get a job. Our companies will go out of business. We'll never be part of the new economy. *Fear.*

Socially, we're told that if we don't keep pace we'll be ostracized, isolated, out of touch. That we won't be able to communicate. That we won't be cool. *Fear.*

Advertising tells us that unless we own all the latest gadgets and devices, we'll never achieve ongoing personal happiness. *Fear.*

The school down the street is giving an electronic tablet to each student. If we don't allow our children to have one, we worry that they may never measure up. *Fear.*

The church on the corner starts putting Bible verses up on a screen during the sermon, concerned that if they do not adapt to the latest trends, no one will pay attention. *Fear.*

People constantly post pictures of themselves on social media, striving to create a viable online persona that will make their lives seem a little less insignificant. *Fear.*

People who are spiritually in tune, however, do not live lives of fear. They do not need to be conformed to the technological imperative of this world, because they have been transformed by the renewing of their minds. They understand that our basic problems stem from our underlying human and spiritual condition, not from which tools we feel compelled to use. Perfect faith is the opposite of fear, and where there is no fear, there is no need for any of its technological overcompensations.

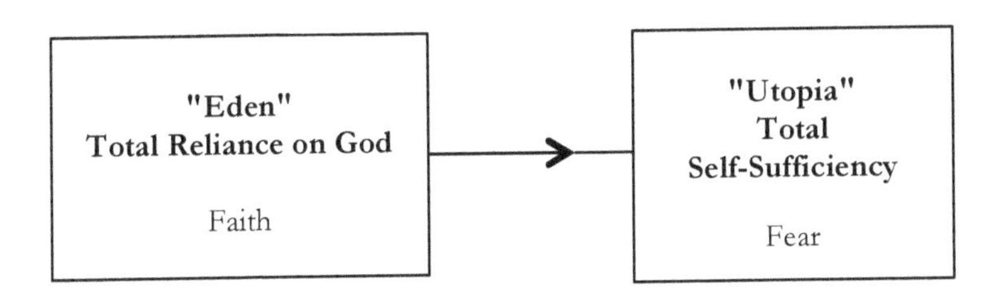

One of greatest spiritual dangers is pride. Conversely, one of the defining characteristics of true spirituality is always humility. This is because, for a spiritually minded person, God is always bigger than me. So is the universe. And humanity as a whole. Nature is bigger than me. Wisdom is bigger than me.

This world was here long before I came into existence, and it will be here long after I am gone. My insignificance in the larger scheme of things is *epic*. However you want to say it—it is not all about me.

Humility has never been high on the technological agenda, however. In *The Garden of Eden*, the snake's offer is that *you can be like God*. The tech mindset takes this offer and runs with it. A recent study showed that the most successful tech CEOs were people who exhibited the following traits: dominance, grandiosity, a sense of entitlement, a lack of empathy, narcissism, manipulation, impulsivity, and a disdain for outsiders.[3] Noticeably absent from the list? Humility.

The ancient Greeks called this *hubris*: the exaggerated sense of self-pride that defies norms of behavior and challenges the gods. It should come as no surprise that hubris runs rampant in the tech world, since whether one is a CEO or not, the very mandate of technology is to replace the works of God with one's own creations. When this is reinforced by the constant propaganda about how technology is making the world a better place for all humanity, it is simply too much to ask anyone involved in the technological agenda to be able to keep a realistic sense of perspective.

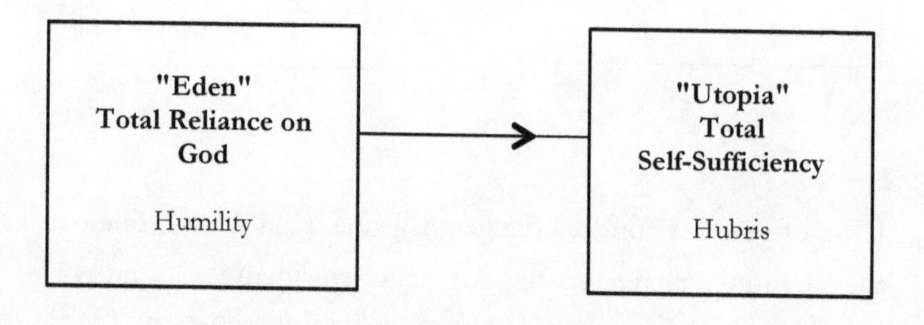

[3] O'Reilly, Charles, et al.

You would think we would have figured out by now that creating a Utopia is just not humanly possible. Starting with the archetypal story of the Tower of Babel right through to the present day, all our attempts to build various heavens on earth have fallen flat. Yet we soldier on, travelling down the continuum away from Eden. Along the way, we are losing the things that *The Garden of Eden* says are good for us: communion with nature, each other, God, and our freedom. In the following chapters, we'll take a closer look at each of these ideas individually, since each merits its own discussion.

I got my kid a direct line to ISIS and the KKK, plus a big stack of X-rated DVDs!
I saved time and just got mine an iPhone!
DOL★FINS
SWIM DAD

"EFFICIENCY"

FOUR IMPORTANT QUESTIONS

Imagine a small time farmer living outside a village in the Middle Ages, working his field with his horse and simple plow. It is quite likely that he would have lived his entire life seeing no new technological developments at all. Changes that did come would have come slowly and incrementally, and been easily absorbed into the existing fabric of life. Those days are long gone. The pace of technological change has increased dramatically. Something seems wrong to us somehow, things we once valued seem to be slipping away, but we don't know how or why. We just feel it in our bones.

Swept up in this movement, we forget that we are responsible for the choices we make. On those rare occasions when we do take an honest look at how much time we spend on our phones, watch television, or play games, the results are usually pretty startling. If we take an even deeper look and examine how these habits are affecting our relationships, it can become very troubling indeed. Then when we dare to come face to face with the degree to which we are dependent on and addicted to the machines we use, we finally start to see just how far we have sunk. Thankfully, since technology offers us a never-ending stream of both hallucinogenic and sedating experiences that can keep us perpetually distracted, we really don't ever need to think about these things at all. It's far easier to just acquiesce.

In this book, however, we are going to delve. The questions will fall into four categories:

- How does technology affect our relationship with nature?
- How does technology affect our relationships with each other?
- How does technology affect our spirituality? and finally,
- How does technology affect our freedom?

We'll look at each, one at a time.

PYRRHO OF ELIS

HOST: Before we start discussing these questions, however, it seems we have a caller on the line. This is amazing, folks! My screener is telling me it's Pyrrho of Elis, the father of Greek skepticism. Pyrrho, is that you?

PYRRHO: Yeah. Thanks for talking my call. Long time listener, first time caller. This new-fangled phone stuff is all Greek to me, LOL!

HOST: How is this even possible? You've been dead a long time.

PYRRHO: Look—you didn't actually think we were going to let you get through this entire discussion without having to answer any skeptical questions, did you? And who better to start things off than me?

HOST: Who do you mean by *we*? Are there any more of you old skeptics around?

PYRRHO: Don't worry, you'll see soon enough. Can I ask my question now?

HOST: I guess so. Fire away.

PYRRHO: I'm not sure I'm entirely clear on something you've been saying. You keep talking about things like plows, and how all technological use has spiritual implications. Isn't this just a matter of semantics? After all, just because you call a plow "spiritual" doesn't mean it's actually spiritual. I looked up the definition of a plow, and it says:

> Plow. *n. A farm implement consisting of a heavy blade at the end of a beam, usually hitched to a draft team or motor vehicle and used for breaking up soil and cutting furrows in preparation for sowing.*

See? There is nothing spiritual about it. A plow is just a plow. Just like an airplane is just an airplane and a computer is just a computer. Technology isn't spiritual at all. It's completely value neutral—it's how you use it that matters.

HOST: Thanks for your question. It's a great one because it illustrates one of the most common misconceptions that people have about technology—that it is value neutral. On the one hand, you are absolutely correct. A plow *is* just a plow. It is made of metal. It has no animate qualities. There is nothing spiritual about it at all. Technological objects are just that—objects. They can be used for good or for evil.

But that's not the whole story. It's like saying that the fruit from the Tree of Knowledge was just a piece of fruit. Technological objects may *seem* value neutral, but they really are not. Neither is technology as a whole. Remember that the snake's effectiveness in the garden comes from telling the truth, but that the truth is mixed with a lie. That's the real semantic issue here.

An unused plow sitting in the corner of a barn is just what you said it is—a hunk of metal that has no spiritual implications. Inherent in its very nature, however, is the idea of its use. That's what makes it a technology in the first place. The second you pick it up and start using it, therefore, you begin to fulfill its intended purpose. In a very real sense we can say that a plow sitting in the corner of the barn is not even fully a plow until it gets used. It doesn't matter whether you use it for good or for evil. A plow *plows.*

That is why static definitions for technology or technological objects are never adequate. While true, they only tell part of the story. What they leave out is the fact that the movement towards self-reliance is *inherent* in the very nature of technology itself. This is true even with something as simple as a plow.

Technology is never value neutral. It's not just a matter of how you use it. It's not that the idea is wrong, but that it's incomplete. To illustrate this further, let's cut a little closer to the bone. After all, it's hard to get excited about plows. Let's talk about guns instead.

PYRRHO: Okay. Shoot.

HOST: —-

PYRRHO: Get it? Shoot? You said "guns", so I said . . .

HOST: Yeah, I got it. Believe me, I got it. Hilarious. On the one hand, a gun is, like a plow, just an object. Here is its dictionary definition:

> Gun. *n. A weapon consisting of a metal tube from which a projectile is fired at high velocity.*

Again—there's seemingly nothing spiritual here at all. A gun is just a metallic object. It's only how you use it that matters. Sure, guns have been used to kill, but it's not their fault. As the saying goes, guns don't kill people, people do. Beyond a doubt, this is true. It is not, however, the entire story.

The question to ask in this context is why guns were ever invented in the first place. Their very creation is an acknowledgment that we are living east of Eden. If we lived in communion with nature, each other and God, we would never have needed them. There would be no reason to hunt, defend ourselves, or go to war. Seen in the light of the continuum, then, a gun *means* self-sufficiency.

The same is true for all technologies: they all move the arrow down the line. This makes it difficult for the person who wants to believe that other people's guns are evil but that his or her smartphone is somehow value neutral. It isn't. This is not to say that all technologies work in the same ways or to the same degree, but in this specific case, which should we fear more? One may be able to kill our bodies, but the other can rob us of our souls. Eventually, I suppose, someone will invent a phone that also shoots bullets. Then where will we be?

PYRRHO: If this is true, then why aren't you modern people having debates about smartphones, along with all the discussions about guns?

HOST: We are. It's just that guns have had about a thousand-year head start. We're just beginning to understand the invasive nature of our electronic devices.

PYRRHO: You know, despite all this, I'm not worried.

HOST: Why not?

PYRRHO: We don't have guns in ancient Elis. Or smartphones. Just plows.

HOST: They'll be coming, eventually. If you stick around long enough.

PYRRHO: I'm not going anywhere. I'm the father of Greek skepticism, after all. And besides, we're all going to have to keep you on your toes.

HOST: Thanks for the call, and thanks for the heads up.

Pyrrho of Elis

QUESTION ONE

HOW DOES TECHNOLOGY AFFECT OUR RELATIONSHIP WITH NATURE?

This is the easiest of the four questions to deal with because everybody is always talking about it. All the major environmental issues of our time, such as deforestation, the extinction of animal species, the use of fossil fuels, etc. are somehow related to our technological usage. The current debate over global warming centers on it. Throughout human history we have moved from a high level of dependency on nature to an ever-growing degree of mastery over it. While this has enabled us to thrive in many ways, as we have moved down the line of the continuum, we have done a tremendous amount of harm in the process. All of this has already been discussed at great length by people who are far more knowledgeable about it than I am.

Let's discuss our relationship with nature on an individual level instead. One of our biggest problems is that technology has enabled us to stop going outside. True communion requires presence. The trouble is, I can now go through the day and have very little contact with nature at all. As I stand in my family room, for example, there are six layers between my feet and the ground: my socks, my shoes, the carpet, the floor, the sub-floor and the foundation. To see the sky, I have to look up though the paint, the texturing, the sheetrock, the ceiling, the rafters, the insulation, the roof, and the roofing tiles. There are several artificial suns in the room for light, and one for heat. Unless a hurricane knocks the house down, I never have to feel the wind or rain. Even if I do go outside, I can walk continuously for thousands of miles on sidewalks or pavement. The only wild animals I generally see are the squirrels that run along our backyard fence. I can literally go for weeks without ever having to look up at the sky or touch the ground.

To say that I am alienated from nature is an understatement. Why does this matter? Because as we have seen, communion with nature is inherently good for us. For millennia, everything we did was tied to our direct experience with the natural world. From the rising and setting of the sun, to the cycle of the seasons, from birth and life to aging and death, we took our timing cues from nature and its patterns. Then, in the last couple hundred years or so, we, not nature, began to determine the tempo. The trouble is, we are still natural creatures who are psychologically adapted to nature's slow, unfettered pace. The technological ethos we have adopted believes that speed and efficiency are always better than going slowly—but this is not true if the rate of change has outstripped our ability to handle it.

We have paid the price for our separation from nature with our mental health. Studies now show that having direct contact with the natural world has a beneficial effect on Alzheimer's disease, dementia, asthma, respiratory disorders, cognition, depression, heart health, hospital recovery rates, obesity, PTSD, stroke, diabetes, ADD/ADHD, autism, stress, brain functioning and overall physical and psychological well being, just to name a few.[4] The ancients, and even our own great-great grandparents, would not have needed any studies to tell them this. They lived in nature constantly. These days, our twenty-four hour cubicles may make us more economically productive, but if I need an alarm clock to wake me up in the morning, espresso to get me through the day and pills to help me sleep at night, something has clearly gone awry.

Another reason this is so important is that when we are no longer in direct contact with nature, it is easy to forget it is there. This makes us far more likely to be neglectful or even abusive of it. Still another layer of alienation is added when our experience of the world is filtered through machines. The person who checks his phone a hundred times a day is the person who has largely forgotten what nature feels like.

Nature cannot be experienced through a device. As I am writing, for example, a super moon is looming outside in the night sky. This is when a full moon coincides with the closest approach the moon makes to the earth, making it look larger than at any other time in its orbit. This is the largest super moon in sixty-nine years, a truly spectacular occasion.

In ancient times, events like the super moon might have been seen as some kind of omen, perhaps to indicate that a famine was coming or that a new child would be born to the king. We know better than that now. The moon is just a hunk of rock to us. We don't go in for all of that superstitious nonsense. What do we do instead? NASA broadcasts a live video feed of the super moon, streaming it over the internet so it can appear on everyone's phone. Compressed down to four inches tall, its awesomeness is gone. We mock the

---

[4] https://www.asla.org/healthbenefitsofnature.aspx

ancients for overinflating its significance, while we squeeze it down to a manageable size and place it between cat videos and *People's Sexiest Man Alive* on some web browser and rob it completely of its wonder. And this is progress? Of course there were some people this week who just went outside and looked up at it. The crazy Luddite anarchists.

Towards the end of *Fahrenheit 451*, by Ray Bradbury, Granger, the leader of the refugees, talks about the influence his grandfather had on his life. His grandfather, he says:

> "hoped that someday our cities would open up more and let the green and the land and the wilderness in more, to remind people that we're allotted a little space on earth and that we survive in that wilderness that can take back what it has given, as easily as blowing its breath on us or sending the sea to tell us we're not so big. When we forget how close the wilderness is in the night, my grandpa said, someday it will come in and get us, for we will have forgotten how terrible and real it can be."

We have forgotten the wilderness because we love technology more. We don't like to admit that this is true, but our actions speak louder than our words. We say we care about nature, but only want to experience it through the very machines that are helping to destroy it. The bigger the screens get and the more powerful the simulations become, the greater our estrangement will be.

Our alienation from nature is a direct result of the self-sufficient impulse we have been discussing. We think we know what is good for us, but as always, our efforts spin out of our control. There is no way the person who planted the first deliberate seed in the ground thousands of years ago could have foreseen the abuses of Big Agriculture. But here we are. As we move further and further down the line, the consequences are inevitable.

While the solution to worldwide ecological problems is difficult, the key for the individual person who is seeking better communion with nature is thankfully much simpler. In the end, it is all a matter of the will. You may never make it back to Eden, but you can still go for a walk. Or look up at the sky. Or taste the rain. Try spending an hour a day outside, just looking and listening; it will transform your life. Just make sure to leave your devices behind you when you go.

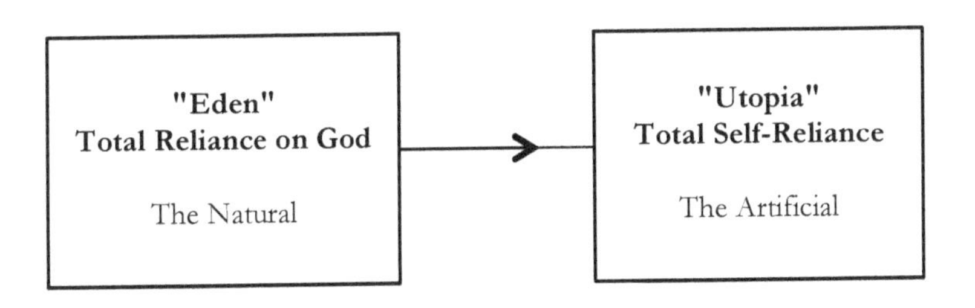

QUESTION TWO

HOW DOES TECHNOLOGY AFFECT OUR RELATIONSHIPS WITH EACH OTHER?

Communication is not Communion

One of the great promises the tech world makes is that it will increase our ability to communicate. We can now reach out and touch someone—instantaneously—from virtually any corner of the globe, and even from outer space. It really is quite remarkable. The claim, however, is much like the lie the snake tells in the *Garden of Eden*—it does not tell the whole story of what is happening: while we are increasing our ability to communicate, we are losing true communion with each other.

We've been staying away from static definitions of words so far, but in this case they are instructive. *Communication* is defined as: 1. an act of transmitting; 2. information transmitted; 3. a process by which information is exchanged; 4. a system for transmitting or exchanging information; 5. a technique for expressing ideas effectively; 6. the technology of the transmission of information.

Communication: A transmission . . . a process . . . a system . . . a technique . . . a technology. These are the words of machines.

Contrast that with the definition of *Communion*: 1. an act

or instance of sharing; 2. a Christian sacrament using bread and wine to symbolize and/or realize the spiritual union between Christ and the communicant; 3. intimate fellowship or rapport; 4. a body of people having a common faith or discipline.

> *Communion: A sharing . . . a union . . . an intimate fellowship . . . a rapport . . . a body . . . a common faith.* These are the words of human beings.

Communication does not necessarily imply relationship. Communion does. While machines are good at enabling people to communicate, they will never produce communion, and communion is what our souls need.

This is why incessant calls for greater communication will never solve our social ills. During my lifetime, for example, there has been ongoing tension between the police and the people in our inner cities. This tension generally stays just below the surface, often even for years, until a specific incident lights the spark that explodes into riots in the streets. This always leads to a lot of belated soul-searching, and inevitably the calls go out for greater communication between law enforcement and the people. Meetings are held, speeches are given, and things calm down. Except that they don't. Time passes by and the cycle repeats itself over and over again.

The problem is not a lack of communication, but a lack of communion. If two opposing groups come together and just transmit information at each other—if they merely communicate—nothing will ever change. If the goal were to achieve intimate fellowship instead, a whole different approach would be necessary. Meetings and speeches would not suffice. Establishing true communion would involve a tremendous amount of time—which we seem unwilling to spend, and a tremendous amount of will—which we don't seem to have. The focus on communication helps ensure that these problems will never be solved.

True communion takes both time and will. Technology, which always seeks the quickest, most efficient pathway to any end, is antithetical to both.

True intimacy and friendship often develop during life's unscheduled moments. It is the spontaneous laugh, the intuitive gesture, the unexpected smile, the times together when no words are necessary that cultivate true communion. These get lost today in the constant cacophony of digital distractions. Buzzes and beeps and vibrations and ring tones require immediate attention and take our focus away from the person in front of us. The unspoken message we send others is that tending to *my* immediate distractions, however trivial, is more important than *our* time together.

This constant focus on the trivial makes people less willing to share their inner lives. Self-disclosure requires self-knowledge, which requires deep thinking. The mind that is perpetually distracted seldom goes there, particularly when the vast majority of the interrupting messages are surface level and trivial.

Is it any wonder then that we have such a societal compulsion to build up false internet personas and create avatars as stand-ins for the true selves we don't even know, much less want to share with others? It is easier to hide behind a digital mask than it is to be vulnerable.

Without vulnerability, there can be no empathy. To truly understand other people's feelings, we need to be in their presence. The non-verbal cues we send to each other, such as eye contact, body language, touch, smell, the sharing of pheromones and the responsiveness to our shared environments all create powerful subtexts to what our words may or may not be saying. To truly understand others, we must be with them.

Consider all the things that *don't* happen when a person texts, for example. There is no body language, no touch, no eye contact. If texting is the primary mode of communication in a relationship, the people involved will never achieve true communion. If the two people can't even be bothered to write out entire words, and if getting together to talk face-to-face is just too much trouble, their dedication to the relationship is suspect. As a replacement for true communion, texting fails miserably.

Another of the tech world's cheap substitutes for communion is what is known as a "shared virtual world", where the mind is tricked into believing that both we and another "person" are really there. Wherever "there" is. By entering into this "world" we can "do" things that are not otherwise possible, and can share "common experiences" with them. Why the deceptive language? To hide the fact that in the real world you can have true communion with another person any time you like, for free. You don't even need to buy any goofy goggles.

The importance of actual physical presence became manifest in the cases of Eastern European orphans back in the late 1980s and early 1990s. When Communism fell, there were thousands of children raised in Soviet institutions that were available for adoption. Many of these children had not been held when they were babies and had been deprived of the normal, loving interactions that occur daily between parents and children. Sadly, many of these children developed severe attachment and behavioral issues as a result, catching many well-intentioned adoptive parents by surprise. Their problems often persisted into adulthood, despite multiple interventions on their behalf. There is a lesson here for all modern-day parents who think they need to buy the latest high-tech-multifunction baby monitor and find a S.T.E.M. preschool for their children. These are not the droids you are looking for. Hugs will do just fine.

That the tech world wants to sell us expensive machines that mediate all of our communications is common knowledge, but the issue goes deeper than mere economics. The motivation is often also personal. It should come as no surprise to anyone by now, but it turns out that the old stereotype of the computer nerd who is socially awkward actually has a strong basis in fact.[5] The very qualities that make someone good at computing are often the same ones that make communion with others difficult. On the positive side, these include being able to block out external distractions, focus for long periods of time on arcane subjects, and work until perfection is achieved. On the other hand, the symptoms also include a difficulty understanding non-verbal cues, a lack of empathy and a general social awkwardness that can be off-putting to others.[6]

This brings up a very important question. We wouldn't ask someone who was colorblind to teach a class on watercolor painting. We wouldn't want someone who was

---

[5] Silberman, Steve.

[6] Mayor, Tracy.

bad at math to determine our economic policy. Techies thrive in environments where they can focus on solving problems "rather than satisfy(ing) the social or emotional needs of others."[7] Why should they, then, define the social parameters for everyone else? It is a task they are particularly ill suited for. And yet, that is exactly what is happening on the internet. Consider, for example, the idea of a social media "friend".

The digital world is the only place where you can be considered a friend with someone even though all you actually do is sit in a room and interface with a box. And yet the word "friend" is still used. This represents a radical redefining of the word. The shift is emblematic of the deceptive use of language that is so crucial to the entire tech enterprise. The techies could, of course, invent whole new words in technologese to describe new phenomena—and sometimes they do. To say that you "Skyped" your mom, for example, is actually quite honest, since the new word was coined to describe the new technological behavior. Too often, however, familiar words are just co-opted in order to mask technology's actual effects.

When it comes to words like "friend" and being "liked" the issue can be very sensitive. Let's say that instead of co-opting the word "friend", the techies had coined a new word, *homie•tech*, to describe the person at the other end of an internet communication, and *fav* to describe the emotional attachment to such a person. "My friend likes me" would have become something like *My homie•tech favs me* in technologese. While this would have been more honest, it would not have been very popular. Creating new words like these would have defeated one of the main reasons for co-opting the language

---

[7] Ibid.

in the first place. If someone had gone around boasting that she had 600 online *homie•techs*, people would have immediately suspected that nobody in the real world actually *faved* her. So the techies co-opted the word "friend" and hoped they'd get away with the bait and switch. Which they have, for the most part.

If we, however, believe that internet "friendship" and true communion are the same thing, then we, being all the more deceived, will never have it.

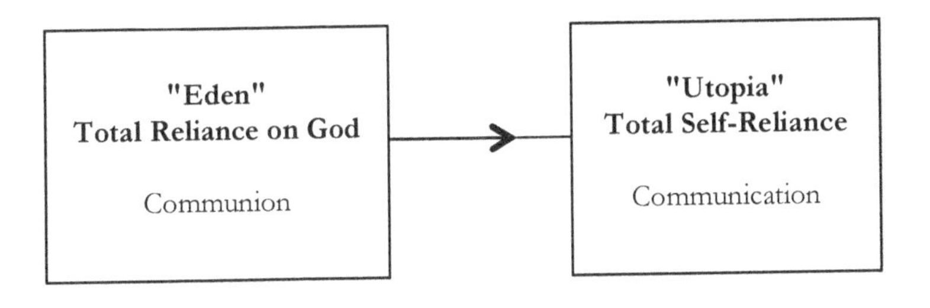

Then

Now

"PERSPECTIVE"

SEXTUS EMPIRICUS

HOST: Before we go on to our third question, I understand we have another ancient Greek skeptic on the line. This time it's Sextus Empiricus. Hi Sextus! I'm thrilled to hear from you, but I'm not so sure our advertising department is.

SEXTUS: Why not?

HOST: We talked with Pyrrho earlier. You old philosopher-types are way outside our target demographic. Besides, you're all too non-materialistic.

SEXTUS: Not Epicurus.

HOST: That's true, not Epicurus. But when was the last time *you* bought anything?

SEXTUS: Well, I got a new robe a couple thousand years ago.

HOST: I rest my case. Do you have a question for me?

SEXTUS: Yes. I'm skeptical about something from the last chapter. You said that people who communicate via text couldn't have true communion with each other. Isn't communion in the eye of the beholder? Aren't you, who grew up without texting, just too old to get it? Young people don't seem to mind.

HOST: First of all, there's something highly ironic about *you* calling *me* old, but you're right about one thing—young people don't seem to mind. That's one of the reasons why I'm so concerned. They've been handed these machines and told to use them, without being given any warnings about their possible side effects. Then they look around and see a generation of adults just as addicted to the machines as they are. It all seems so normal. Why should they think any differently?

SEXTUS: That all may be true, but that wasn't really my question. What about texting and the impossibility of true communion?

HOST: The issue is not the nature of texting itself—it's the blurred line between communication and communion. If people were able to limit their technological interactions to the occasional processing of mundane bits of information, there wouldn't be much of a problem. But they can't, due to the ever-progressing nature of technology. Remember that technology is a movement that has implications for our relationships. People are now using machines as their primary means of social discourse. As a result, they are losing the ability to interact face to face.

Let me give you an example. I was recently at a coffee shop talking with a friend. Four people were at the little table next to ours. All of them were *deep* into their phones the entire time. They rarely talked and for the most part never looked at each other. Every so often, one of them would say something like, "I need a napkin" or "it's kinda hot in here", and another would answer "ok" or "uh huh" without bothering to look up. This went on for *forty-five* minutes. They had taken the intentional step of going all the way to a coffee shop to sit together, setting themselves up beautifully for the opportunity to have true communion, but their technological obsessions and addictions overpowered them instead. At one point, one of them got a call. Evidently someone on the other end asked what he was doing. "Oh nothing, he said, "we're just sitting here at Starbucks, talking." *Talking.* I'm sure he actually believed it.

SEXTUS: C'mon, aren't you being a little harsh? Aren't those just your definitions of communication and communion? Who are you to impose your definitions of words on others?

HOST: I'm not the one imposing my definitions of words on others. I'm using the old ones—the ones that were around long before I was born. It's the technological world that is distorting the meanings of words. It wants people to believe that they are achieving communion when what they are actually doing is merely communicating.

SEXTUS: Why would they want that?

HOST: Two primary reasons. The first and most obvious is money. They want us to be customers for life, so the machines are purposefully made to be addictive. Once we are ensnared, we need a way to justify our compulsive behavior. This is easier to do if we *think* we are getting true communion as a trade-off for our addictions. Hence the massive propaganda effort directed at getting us to believe it.

The second reason is something I alluded to in the last chapter. It's what I call the "ordination of limitation"—the universal human desire to recreate the world in our own image and to regulate the behavior of others in accordance with our own deficiencies. It's something we all wish we had the power to do at times. For techies, it works like this: since they are more comfortable interacting via machines, wouldn't the world be a better place if everyone else had to as well?

Before, after all, it was always some popular kid who hosted the Friday night keg parties at his house. People always got left out. Now, the party's online, with technology mediating *all* the social interactions. It's a huge ego rush. Of course you have to change the meanings of words like "host" and "party", but that doesn't matter. Co-opting older words is just an inevitable part of the social transformation.[8]

SEXTUS: This sounds a lot like *Revenge of the Nerds.*

HOST: How does an ancient Greek skeptic know about *Revenge of the Nerds*?

SEXTUS: A bunch of us old philosophers formed a movie club. We watch movies together all the time. Plato, for example, loves *Legally Blonde,* and has seen it fourteen times. He says Elle is his ideal woman.

[8] Another example is "surfing", as in "I'm surfing the net." What could possibly be cooler than that?

HOST: Wow. *Plato.* Who could have imagined? Actually, *Revenge of the Nerds* is the exact opposite of what I'm taking about. The film is charming because it shows how the nerds have a legitimate place in the existing social structure. Who can argue with that? The fact that the existing social structure in the film is the product of an elitist fraternity system makes us root for them all the more. In the real world, however, technology does not just want to fit in—it wants to take over and *become* the elitist system itself.

In the film, the nerds don't assume their place in the world until they are able to overcome their fears and confront their tormentors face-to-face. Techies now want a world where everyone interacts solely through machines—one where there is no need for face-to-face interaction at all.

This differentiates them from ordinary introverts, who have very little interest in telling other people how to live. The artist and musician Prince may have holed himself up in Paisley Park, for example, but he never tried to force everyone else to wear purple.

It also separates them from many of their technological predecessors. In the early days of computers, techies weren't out to rule the social world. They just went about their day-to-day business like everyone else. With the advent of the personal computer and then the creation of the internet and social media all of this changed, as the sort of domineering influence we see today finally became possible.

SEXTUS: Are you suggesting that all techies today are out for social domination?

HOST: Of course not. But a certain technological *mindset* is. Many of the people working at the tech companies today have real, legitimate concerns about the way things are going. The computer labs at our schools are staffed by long-suffering saints just trying to help teachers and students stay afloat. But they are seldom the ones driving the bus. They're being taken along for the ride just like the rest of us.

SEXTUS: But isn't this just a matter of societal norms? Who's to say what normal communication is?

HOST: It's not about what is normal. It's about what is inherently good for us. A society where people no longer look each other in the eye is one where there is no longer any true communion. It would be devastating to our humanity and our spirituality.

SEXTUS: But the tech gurus are always claiming that their machines are bringing the world together. It's one of their main selling points.

HOST: It's just another deceptive use of language. To say that something like texting is bringing us together is to change the ordinary meanings of the words. Since we can text without moving anything except our thumbs, at no time does it "bring" us "together" anywhere. If anything, the machines actually create an even greater space between us.

SEXTUS: I agree that this might be a problem if all someone did was text, but nobody does that. Texting is just one facet of an exciting new tableau of communication strategies that are now available to today's digital citizen of the world.

HOST: Are you kidding me? *Tableau?* Did you come up with that yourself?

SEXTUS: No, I read it somewhere in a brochure.

HOST: That proves my point. Nobody does just text. They use all the other *facets,* as well, which leads to an ever-increasing proportion of their interactions being mediated by machines.

SEXTUS: Are you saying that texting is useless?

HOST: Not at all. It certainly does do *something.* It just doesn't bring us together in true communion. At the very least, techies ought to be honest and invent new words that accurately match the new realities, not co-opt older ones in order to mask their intentions or overinflate the importance of their machines. As George Orwell said, "The great enemy of clear language is insincerity."[9] Even "text" itself is a co-opted word.

SEXTUS: I have one last question. So what? Who cares if the techies change the meanings of words? Isn't language always evolving?

HOST: Yes, but it's important to understand the nature of each change in order to not be manipulated by it. Not all language changes are innocuous—some are purposefully designed to be deceptive. A generation that hasn't experienced true communion, for example, will never know what it has missed, even if it continues to use the word.

---

[9] *Politics and the English Language.*

In nature, an apple is still an "apple" and a cloud is still a "cloud". In the world of people, communion is still "communion", and is still inherently good for us.

Sextus Empiricus

QUESTION THREE

HOW DOES TECHNOLOGY AFFECT OUR SPIRITUALITY?

Technology is a jealous god that will permit no other gods before it. Since it is a movement, it is appeased only when we are going in the direction it wants to take us. Since it affects relationships, it wants to mediate all of our experiences with God. Traditionally, people have expressed their spirituality through a) contemplating nature; b) listening to the Spirit, meditation and prayer; c) reading scriptures; and d) meeting and worshipping together. Technology is now actively making it harder for us to experience all of them, and is offering itself up as a replacement instead. For people of faith and spirituality, the time has come to decide whether or not we want to give it the keys to the kingdom.

We've already discussed communion with nature, so let's look at the other expressions of spirituality one by one:

LISTENING TO THE SPIRIT, MEDITATION & PRAYER

Listening to the Spirit, meditation and prayer all require non-distracted quiet. This has always been hard to find. Now that people look at their phones over one hundred times a day on top of all their other distractions, it has become almost impossible. The machines have robbed us of the solitude necessary for listening.

A story from the Old Testament is instructive in this regard. Elijah is told to go and stand on the mountain, "for the Lord is about to pass by".[10] A great and powerful wind tears at the mountain and shatters the rock. God is not in the wind. After the wind comes an earthquake that shakes the ground. God is not in the earthquake. After the earthquake comes a great fire. God is not in the fire. Finally, the voice of the Lord comes in the still small voice of a gentle whisper.

God does not shout. He does not force himself upon us. He is there, and will speak if we listen, but he will not insist. Technology is now working hard to make sure we never hear him.

Meditation requires a tremendous amount of self-discipline and focus. Technology, on the other hand, promises instant gratification. No discipline is necessary to jump from screen to screen, site to site, distraction to distraction. It doesn't matter which tradition you follow: if you are constantly tethered to your device, you will never:

- Pray without ceasing (Christianity)
- Meditate on God's word day and night (Judaism)
- Contemplate a koan (Zen)
- Achieve Samadhi through Dhyana (Hinduism)
- Engage in Simran (Sikhism)
- Engage in Dhikr (Islam)
- Achieve enlightenment (Buddhism)

---

[10] I Kings 19:11-12

The technological counterfeits to true spirituality are everywhere, no discipline required. People now get messages on their phones reminding them when and how to meditate. Headsets measure electrical brain waves and alert wearers when their minds wander, allowing them to return to meditation more efficiently. A VR app features psychedelic purple graphics and a simulation of the bodhi tree that claims to provide a "journey to enlightenment" in twenty minutes. Machines are in the works that will be able to instantaneously stimulate our brains by mimicking the effects of meditation as experienced by Tibetan monks.

It's all a sham. The person who meditates on God's word day and night goes through a process. Achieving enlightenment takes time. Technology may be able to instantly simulate brain functions, but that is no substitute for the real thing. The time, the effort, the discipline, the difficulty and most importantly, the failures along the way are all part of the process toward true spiritual maturity.

Failure is one of life's greatest teachers. If I forget to meditate or pray, my very act of forgetfulness betrays an honest truth—that maybe I'm not so spiritually in tune after all. It's a vital lesson to learn. Having a device there to constantly remind me what to do allows me to live in the illusion that I'm already there, whether I am or not.

Even when the technology eventually gets so powerful that it will be impossible to distinguish the real experience from the counterfeit, the wearer of the headset or goggles will always know one thing when he or she takes them off: that whatever it was that just happened was artificially created by a machine. This is the opposite of spirituality.

Prayer too, is an anathema to technology. Why wait around for an answer from Jehovah or Allah when you have been granted an audience with the great god Google and his consorts Siri and Alexa, who give their answers instantaneously?

God is not interested in efficiency. He is interested in what is good for us. In the Old Testament, he allows the children of Israel to wander forty years in the wilderness—a trip that should have only taken a couple of weeks. It is only after they've painstakingly learned the spiritual lessons that they are supposed to that they are finally allowed to enter the Promised Land. Today, Moses would just buy a GPS device and save himself all the trouble. Would this have represented progress? It all depends on what you are after.

READING SCRIPTURES

Not all cultures have had sacred texts, for the simple reason that not all cultures have had writing. Of those that have, however, *all* have had books that have been set apart as holy.

Let us set aside for a moment the question of whether language can ever fully convey comprehensible thoughts about God, or whether writing, as a technology, can ever impart spiritual content. These are important and interesting questions and are worthy of discussion.[11]

For now, we will say this: if you are among those who believe that certain books such as the Bible, the Qur'an, or the Bhagavad-Gita are holy, you are in trouble. You are even in trouble if you believe that the works of our greatest authors and poets are inspired in some more earthly manner.

[11] The first is beyond the scope of this book, and has been written about elsewhere at great length. The second will be discussed in a later chapter.

None of them will survive the digital age in any meaningful way.

People used to believe that as the internet developed, reading and writing would simply be transferred over to it and that all would continue as it was, without any real change. It didn't happen. People who spend an inordinate amount of time in the image-based, non-linear on-line environment (yes, an oxymoron) literally *cannot* read as well or as easily as those whose primary intellectual landscape is print. [12] This is especially true when it comes to the type of deep meditative reading required by scriptures. Reading a book like the Bible, with its multiplicity of authors, genres, cultures and eras, has always been difficult enough. In the coming years, it will be nearly impossible for people to do.

The first thing that is being lost today is the very idea that a book can be considered sacred in the first place. When I was twelve, I attended my friend Randy's *bar mitzvah.* One of the things that made a deep impression on me was the care that the rabbi took with the Torah. It was taken out of a special cabinet (called the *aron kodesh*) and carried around the entire congregation before going back up to the platform where Randy stood waiting to read. To show their respect, the people kissed their hands and then touched the large scroll as it passed by. As it got closer to me I remember having a brief moment of panic. What was I, as a Baptist kid, supposed to do? I wondered if my underlying instinct for ecumenical unity would be strong enough to overcome any sectarian hesitancy I might have had. (Actually, I didn't wonder this. I was only

[12] See *i-Minds* and *The Shallows* for excellent discussions on the effect screen time has on people's brains.

in the *seventh grade*, for heaven's sake. I was really just afraid of looking stupid—a fate worse than death. I went ahead and kissed my hand and touched the scroll). Randy then read the Hebrew text. I can still hear the words in my mind: *Baruch atah Adonai Eloheinu melech ha'olam.* When he was finished, the Torah was carefully placed back in its cabinet.

The entire ceremony was a paean to literacy. It was reading that marked Randy's passage from childhood to adulthood. It was reading that signified his membership in the community. It was reading that led the voice of God, the years of Jewish tradition and now Randy's life to be somehow mysteriously intertwined. All this was embodied in the Torah, which was therefore regarded with ritual and reverence.

To take the words of the Torah and place them on the internet is to rob them of their distinctiveness. Online, they are no different from billions of other bits of information. It would be like going into a synagogue and seeing a billion different cabinets with a billion different Torahs in them. Or like reading one long continuous scroll whose content jumped seamlessly from politics, gossip, pop culture, pornography, chat rooms, and finally, the Torah, only to be followed by dating services, baseball scores, Susie's vacation pictures, and how to make a banana cream pie, all stretching off to infinity.

Who is to say what is important here? Who has the authority to make the determination? The internet democratizes all information and breaks down all cultural walls of authority. All content becomes one and the same thing. The Torah on your phone is not the same as the Torah in your hand.

The shift that is taking place seems especially difficult for the generations raised on books to understand. It is a matter of perception. Growing up with a printed Bible, you knew that there was something special about it. Even if you didn't believe in it, it still was not like other books, not to mention being different from television, movies, magazines and newspapers. And your phone, of course. It might have even said "Holy" right on the cover. You knew it came with thousands of years of societal sanction, even if you didn't believe in what it said.

So when the Bible was first placed on the internet, the book people all said, "Look—there's our holy book that has been placed online. Isn't that great?" But this was only because they perceived of it as being a book in the first place. Not only that, but as a sacred book. Since it was already set apart from all other books before it went online, they thought it would continue to be set apart once it was.

Things didn't turn out that way. For digital natives, such distinctions are purely arbitrary. The internet is not a book. It doesn't really have "pages"—it's just another word that has been co-opted. Since internet content of all types is accessed in the same ways through the same machines, there is nothing to physically set the Bible (or any other book) apart from anything else. Where everything is equal, nothing is exceptional.

There is no way to tell which of a trillion websites has any authority. For every site with genuine biblical content, there is another just a click away telling you why the Bible is ridiculous, or stupid, or even evil. Which one of the billion scrolls in the digital "synagogue" will you read? Which one should you trust? Why should you believe something just because it once came from a *book*? Aren't books obsolete? When all information becomes equally important, it really means that no information is important at all. All books on the internet are equal, but no books are more equal than others.

It turns out the rabbis had it right. The Torah deserved to be handled with care. It deserved to be stored in the *aron kodesh.* It even deserved to be kissed by an awkward junior higher. Today, it still deserves to be read in a sanctified manner. This is unlikely to happen online.

Every kid who grew up in church remembers developing certain coping mechanisms for dealing with boring sermons. My sister and her friends used to go through the hymnal and add the words "in the bathtub" to the end of each hymn title, resulting in such comedic gems as *O Come, Let Us Worship (in the bathtub), I Sought the Lord (in the bathtub),* and *Rise Up, O Men of God (in the bathtub*), which sent them into giggling convulsions. In addition to drawing pictures on the bulletin when I was bored, I used to browse through the pew Bible just to see if I could find anything interesting. One day, I discovered the book of Habakkuk.

Up to that point in my young life I hadn't even known Habakkuk existed. It turns out that this small, obscure book is pretty interesting. Its declaration that *the righteous shall live by faith* (2:4) is quoted extensively by New Testament authors. Habakkuk himself was one of the few prophets to openly question God's wisdom. I personally get a sardonic chuckle today from the verses that talk about Babylon gathering the nations in a *net* before they are destroyed (1:15-17), even though this is completely out of our context here.

The point is not what Habakkuk is about, however, but how I found it. When I opened the pew Bible, I could see that its contents were laid out in two main sections, called testaments, and that each testament was divided into a number of books, and that the books were thematically arranged. I could see their titles, how many of them there were, and generally how long they were. The name "Habakkuk" caught my eye and I decided to give it a try. While elements like the table of contents were, of course, later additions, the fact that Habakkuk was on the list of biblical books meant that it came with a certain authority and had stood the test of time.

The internet experience is not like this. It is more like outer space—expanding all the time. Its content is, for all practical purposes, limitless. But it is not completely without organization. The tools that have been created to try to make sense of it all are the search engines, such as Google or Bing. Without them, we'd never be able to find our way around. The problem with search engines is that their algorithms are structured so that the most popular hits come up first. The result of an internet search is the list of the sites that people want to see the most. Not the sites that are the truest, or the most authoritative, or the best. The ones that are the most popular.

This creates a problem for Habakkuk. Even in their day, the dour prophets with their predictions of gloom had trouble enough being liked. Placed online, the book now has to compete for popularity with not only John 3:16 and Psalm 23, but a trillion other pieces of non-related information as well. And since we can now search directly for the specific verses we already know and like, Habakkuk is sure to fade further and further into obscurity. Who will ever search for it?

Even more pernicious is the fact that the search engine companies make the vast majority of their money per search. Advertisers buy space where their products are most likely to be seen. *This means that John 3:16 makes Google more money than Habakkuk does.*

Since Google is a business, it has no ethical obligation to keep offering access to content that does not continue to make it any money. Unless searches for Habakkuk continue, you can, for all practical intents, kiss it goodbye. And if no one can find it, what good will it be, even if it is still there? The Bible on your phone is not the same as the Bible in your hand.[13]

"But Google would never do that," you say. "It's a *good* company." Why? Because it has a colorful, childlike logo and its workers get to ride skateboards through the office? And even if the current workers at Google are somehow uniquely altruistic, there's no guarantee that the next generation will be. The fact that Google is now in the process of scanning all the books ever printed into digital form, and that it will then control the access to all of these books, is a totalitarian nightmare just waiting to happen.[14] A librarian friend warned me about this years ago: "Once all information is in the 'cloud'," she said, "whoever controls the 'cloud' will control all the information."

George Orwell famously predicted this in *1984*: *Who controls the past controls the future. Who controls the present controls the past.* I wonder if the average person sitting in the pew on Sunday morning is aware of the implications of looking up individual verses on his or her device: the verses you are searching for today will help determine the verses you will not be able to find tomorrow.

And when Google, or one of its inevitable successors, decides for whatever reason that it doesn't want you to have

---

[13] You may have noticed that the Bible online is almost always surrounded by advertisements of all kinds—an ironic mixed message in light of Jesus's anti-materialistic teachings.

[14] See chapter 8 of *The Shallows.*

access to the Bible at all, it will be gone. Why should we allow a tech company to have this kind of power?

My childhood church, ironically, acted a lot like a search engine in that it effectively narrowed my biblical exposure by default. Over the years, I heard lots of sermons about John 3:16 and exactly none about Habakkuk. A cynic might say this is because the organized church is a business too, and that John 3:16, with its optimistic promise of salvation, puts more tithers in the pews than the books of the prophets do. I would prefer to believe that unlike Google, my church also had my spiritual growth in mind. Who knows? I was just a kid. In either case, my church did place Habakkuk right there in the pews. It was only because I had a physical copy of the Bible in my hand that I was able to stumble upon it in the first place, and was then able to read it for myself. The Bible on your phone is not the same as the Bible in your hand.

I wonder what the kid in the pew this Sunday will be able to find on his smartphone during a boring sermon. . .

According to social critic and media ecologist Neil Postman, being able to read requires certain habits of mind. These include:

- The ability to operate at a high level of abstraction
- The ability to think logically and sequentially
- The ability to maintain a high level of self-control and focus
- The ability to defer gratification
- The ability to take initiative[15]

---

15 *Amusing Ourselves to Death.*

Studies show that when we read, the act creates physical changes to the structures of our brains. It is a lot like any other muscle. The more we read, the stronger we get. This shows up on brain scans, where the brains of readers actually look different than those of non-readers. This extra boost gives readers certain advantages: Not only does reading more make us better readers, it also makes it easier for us to think logically, sequentially and abstractly, to develop self-control, to defer gratification, and to learn how to take initiative, since as we read, we are constantly practicing these habits of mind.

The experience of being online is very different. It is not bound by rules of logic or sequence. Its content mostly consists of visual images, ruling out the need for abstraction. What text there is usually comes in small paragraph length chunks that require very little focus. Unlike words on a page, the text is often colorful and moves and jumps around. It is often accompanied by sound. As a person clicks from link to link, the experience is scattered, random, simplistic, instantaneous, and passive. After even ten minutes, it is nearly impossible to recreate or even remember the sequence of choices that led us to the screen we are currently on.

Because the internet world places so few demands on the mind, the literate brain has very little trouble navigating though it. The person who can read *Hamlet,* for example, needs very little coaching on how to do an internet search. The fact that three and four year-olds are now using iPads is proof that the intellectual challenges here are pretty slim.

Unfortunately, the opposite is not true. The person who is constantly online will simply never develop the mental muscles necessary to be a good reader. In fact, we now know that a steady diet of image-based screen content actually makes it *harder* for a person to read, just like a junk food diet and sedentary lifestyle have an adverse effect on a person's ability to run a marathon.

While the cognitive consequences here are dire, there is spiritual fallout as well. Because of their unique content, scriptures call for a unique type of reading. The experience is spiritual as well as intellectual. It is usually done slowly. There are often long pauses for contemplation and self-examination. Listening is a key part of the process. In Islam, reading the Qur'an is in itself a form of worship. On the internet, however, the constant pop-ups, choices of links to click to, instant messages coming in and the barrage of advertisements make reading like this nearly impossible, even if a scriptural text is up on the screen. Also, the entire time we are trying to read scripture online, we are constantly aware that all kinds of tantalizing, magical worlds are just a click away. True focus is impossible. The Bible on your phone is not the same as the Bible in your hand.

It is not a coincidence that self-control, the ability to focus, the delaying of gratification and the taking of initiative are all marks of a spiritually mature person. By subverting people's ability to read, the digital world is cutting off our access to the scriptures and thus to one of the key pathways to our spiritual growth.

MEETING AND WORSHIPPING TOGETHER

And the Word became flesh and dwelt among us.
John 1:14

Although each person must come to God individually, the spiritual life is meant to be lived in community. Even though he didn't need to, Jesus called twelve disciples to be his constant companions, travelling, working, eating, and joking with him as friends. After his resurrection, they formed the church, which comes from the Greek word *ekklesia* and means "the gathered ones". Christians are called to live out their faith by going into the world and feeding the poor, healing the sick, caring for widows and orphans, welcoming social outcasts and comforting the afflicted.

This is not unique to Christianity. All the great faith traditions in some way incorporate bodily presence together as a key component of both their worship and their service to the world. As anyone who has ever experienced it knows, it is a beautiful thing to be spiritually in tune with others.

> *All the believers continued together in close fellowship and shared their belongings with one another . . . day after day they met as a group in the temple, and they had their meals together in their homes, eating with glad and humble hearts, praising God, and enjoying the good will of all the people.*
>
> *Acts 2:44-47*

This type of unity is not even necessarily unique to groups with religious affiliations. Any group that wants to achieve a true sense of togetherness needs to meet in person. Few understand this better than soldiers who have fought together on the battlefield. Every summer, thousands of young people seeking a communal experience travel to music festivals around the world. For an example of unity that is truly unique, check out the Black Hole at a Raiders football game some Sunday afternoon—if you dare. Cain famously asked, "Am I my brother's keeper?" The world's answer is, resoundingly, yes.

> *Two are better than one*
> *Because they have a good return for their work.*
> *If one falls down*
> *His friend can help him up.*
> *But pity the man who falls and has no one to help him.*
> *If two lie down together, they will keep warm.*
> *But how can one keep warm alone?*
>
> *Ecclesiastes 4:9-11*

Even when true communion has been achieved, however, it has been notoriously difficult to maintain. As I write, there are fifty-six different *types* of Christian churches in my area, twenty-one of them Baptist. Across the globe, Sunnis and Shiites continue the battles they have fought for hundreds of years. Synthetic drugs are needed to help manufacture the communal feeling at the big music festivals. A Dodgers fan attacks a Giants fan in the parking lot after a baseball game, leaving him paralyzed.

Our hunger for community also leaves us open to counterfeits of all kinds: Jim Jones and The People's Temple, the Manson "family", the Crusades, terrorist groups, street gangs, political parties. We are so desperate for unity that we will willingly give up what is good for us in order to have it. Barring an intervention of the Spirit, a splintering process seems to be inevitable.

> *For where two or three come together in my name, there I am with them.*
>
> *Matthew 18:20*

A technological manifestation of this splintering is what is called the "online church". This is where a person sits alone in a room for a period of time interacting with others via a machine, and then claims to have "gone" to church. The physical presence of others is optional. It is one of the saddest of all the ways that technology has counterfeited authentic spirituality, yet its adherents sincerely believe they are doing something good. As is so often true, the anesthetizing effects of technology can keep us from recognizing the deception.

Here are some of the problems with the online church:

1. You have to be able to buy your way in. Participation is limited to those who have purchased the proper consumer products. So much for salvation being free! This is "first world" religion at its worst. And once you have made your initial purchases, you're still not off the hook, because you'll need to continually upgrade your machine and its software in order to keep participating, just as a Scientologist has to shell out more and more money to keep ascending to Clarity. John the Baptist said one of the reasons he knew Jesus was the Son of God was because the Good News was being preached to the poor. Not so with the online church.

2. It's inherently narcissistic. It's already bad enough that there are twenty-one different Baptist churches in my area. With the online church, there can now be a million versions of the Church of Me. That's not to say that there aren't any advantages to membership in the Church of Me. After all, I get to control all the variables. I say where, when and how. I set the room temperature and get to sit in my comfy chair. If there is something I don't like, I can turn the volume down, or the screen off. When I get bored, I can switch to baseball. The online church is all about self-determination—precisely

the same choice the man and woman opted for in *The Garden of Eden.*

3. It involves the same predictable verbal evasions that plague other technologies. To the online church a "meeting" is not a meeting, a "congregation" is not a congregation, a "conversation" is not a conversation, and "together" does not actually mean together. The words strain against their natural boundaries until they finally burst into a flood of incoherence. Anything becomes justifiable as long as I get to define the terms. Once again, the techies should just be honest and create new words to describe the new realities. They won't in this case, however, because to do so would expose the duplicity. The online church is not church, not a congregation, and not communion—until you redefine those terms.

> *And let us consider how we may spur one another on toward love and good deeds. Let us not give up meeting together, as some are in the habit of doing, but let us encourage one another.*
>
> *Hebrews 10:25*

Of course there are those who do gain some benefit from certain on-line ministries. A church might post its service on-line for the benefit of those who are too sick or elderly to attend a service, for example. A person looking for a church to attend might happen upon a church's website. Technology, as always, does do *something.* How much better would it be, however, if the sick and elderly weren't left alone in the first place? And if a church were really doing its job, it wouldn't need to depend on marketing techniques to lure people in. *They shall know us by our love*, not by our websites.

Thankfully, it is also true that God can redeem any of our misguided efforts and use them for good. Consider, for example, the sad case of religious television (which is also, of course, a technology). What may have started out with good intentions quickly became a mass of financial, sexual and spiritual improprieties. Doctrines were created for the express purpose of fleecing innocent victims out of their cash. Even worse, *all cosmetic restraint was abandoned.* For those of us of a certain age, the image of televangelists going down waterslides is indelibly burned into our brains. The chicanery continues to this very day. And yet, despite the damage done, *some* good does come out of it all. For someone. Somewhere. I hope. But the fact that God can save us from ourselves does not mean that we should institutionalize our errors. Should we continue to use technology more so that God's grace may thereby increase? May it never be.

> *I have much to write to you, but I do not want to use paper and ink. Instead I hope to visit you and talk with you face to face, so that our joy may be complete.*
>
> *2 John 1:12*

Finally, a thought experiment: You get to travel back in time to the first century. Somehow, your computer comes with you, and it works. You are given a choice: You can spend a day in the physical presence of Jesus, hanging out with him and the disciples in Galilee, listening to him preach, watching him heal the sick, and finally having him tell you that he loves you, or you can sit alone in a room all day and communicate with him via FaceTime.

To suggest that the two would somehow be equivalent is, well, ridiculous.

We return once again to the continuum:

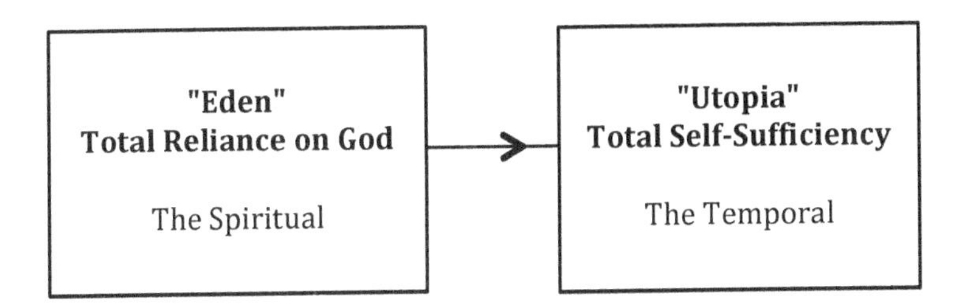

SAINT THOMAS

HOST: Welcome back from reading, everybody. Let's go ahead and take our next caller. Caller, are you there?

THOMAS: Hi! Thanks for taking my call. It's me, Saint Thomas.

HOST: You mean Saint Thomas the disciple? The one known as *doubting* Thomas?

THOMAS: In the flesh.

HOST: Great, another skeptic! How do I know it's really you?

THOMAS: If we were in a room together, I would just let you touch my side. I've always been a stickler for empirical evidence.

HOST: Too bad we have these machines as intermediaries.

THOMAS: Too bad for you—I know who *I* am. You're the one who is going to have to take things on faith.

HOST: So I've read. Do you have a question?

THOMAS: I do. If technology is moving us away from true spirituality, where does that leave the scriptures? Aren't works like the Bible, the Torah and the Qur'an *books*, and aren't books themselves technologies? Are you suggesting that they actually take us away from God?

HOST: No, but that's a great question. You skeptics seem to be full of them. I guess that's why you're skeptics in the first place! You're right: books are technologies. Let's take it even one step further. Writing itself is a technology too.

THOMAS: How so?

HOST: It's a rather long, complex story. There once was a time before writing existed. Walter J. Ong calls this period *Orality*.[16] During that time, all spoken language was live and took place in real time. Once a conversation stopped—that was it. It was over. Anything that people needed to remember had to be stored in their memories. Unless someone was talking to herself, all language was always *relational.*

Writing changed all this, and brought about the age called *Literacy*. The creation of the phonetic alphabet meant that words could be broken down into phonetic parts (phonemes), and that those parts could be represented with abstract symbols (letters), and that those symbols could be reproduced (writing).

For thousands of years after writing was created, the vast majority of people continued to live in Orality. Reading and writing were reserved for the upper classes and elites. It

---

16 *Orality and Literacy.*

wasn't until Gutenberg's invention of the printing press in AD 1440 that literacy started to become widespread, but the change happened slowly. Even now the shift is not quite complete. There are still a few non-literate cultures in the world.

THOMAS: What difference does the shift to Literacy make?

HOST: It means that language no longer has to be relational. Since the words are stored on the page, the reader and writer never have to meet face to face. The words can be written a day, a week, or even centuries before they are read. And since writing preserves language in a fixed form, people no longer have to remember everything themselves.

THOMAS: What do they need to know instead?

HOST: At the very least they have to know how to read. But writing even changes what "knowing" means. A person from a literate culture does not know things in the same sense that a person from an oral culture does. On the one hand, writing radically expands the number of things we can know about, since the information can be stored in print rather than in our heads. On the other hand, it also radically expands the number of things we can forget, since we can always just look them up. In an oral culture, people *really* have to know the things they know.

THOMAS: How do people from oral cultures remember things if they can't write them down?

HOST: They use all sorts of mnemonic devices. These take the form of stories, rituals, ceremonies, music and art. Certain people and objects merit special symbolic significance. Figurative language, like rhyme, meter, alliteration, metaphor and repetition are used to help preserve key ideas.

THOMAS: That sounds a lot like poetry.

HOST: It is. The role of the bard or poet is extremely important in oral cultures. Poetry for them is not just another form of personal expression. It is a vital means of societal cohesion. Interestingly, the more technologically advanced a society gets, the less it values poetry. And story. And folk arts of all kinds. They just aren't needed as much.

THOMAS: What else happens?

HOST: In oral cultures truth is conveyed in different ways than it is in literate ones. If you were to ask a village storyteller from Orality about his tribe's history, your question would not be perceived as a request for abstract information. It would first be perceived in terms of relationship, since in Orality all language is relational. The answer would be tailored to *you*, in particular, in *this* moment, and would just as likely come in the form of a parable, a poem, or even a dance. This would not mean that the answer was not true, but just that the storyteller might have thought that a poem was the best way to reach his intended relational end with you. To those of us raised in a literate culture, this type of answer can be bewildering.

THOMAS: Why?

HOST: Because we're used to the world of literacy. We don't really have story *tellers* anymore. With writing, language can be analyzed. Words can be read again and again, checked for logic, and are subject to interpretation and correction. Being accurate becomes more important than being relational, because the words are separated from the people who write them. Today, we would consider figurative language an inappropriate way to answer a historical question, and would be more comfortable with expositional prose.

THOMAS: Okay, but where does all this leave the scriptures?

HOST: In a rather unique spot. Let's use the Bible as our example. It is a collection of books, so it is a product of Literacy, but the vast majority of its content emerged out of Orality first. This makes it an interesting kind of hybrid. Take Genesis, for example. There were no eyewitnesses at the creation, obviously. The story was first told orally and then passed from generation to generation before it was finally written down. When it was, the first chapter took the form of a poem with a refrain, which is an oral form. So what does that make it? A product of Orality, or a product of Literacy?

This duality continues through the rest of the Bible. The book of *Proverbs* is a collection of wise sayings, and *Psalms* is a collection of songs that had been sung. Orality. The Ten Commandments, conversely, were written down first, on tablets. Literacy. In the New Testament, Jesus was highly literate, but he never wrote down the things he said, and he often told stories in the form of parables, an oral form. What are we to make then, of the Gospels, where his words were collected and written down by his followers? The letters of Paul, on the other hand, were written first, and clearly display the type of linear, analytical thinking typical of a mind trained in literacy, even though his intent was always highly relational.

So what is the Bible? Is it a product of Orality whose primary purpose is the relationship between the speaker and listener? Or is it an abstract, literate text that is meant to be analyzed? Or some kind of combination of the two? Does taking a story from Orality and technologizing it by putting it into writing change its essential nature? If so, would this shift move us down the continuum away from God? These are important questions.

THOMAS: Uh, I believe that last one *was* my original question, wasn't it?

HOST: So?

THOMAS: So it means you've been talking now for almost four pages and still haven't answered it.

HOST: That doesn't make any sense. One does not *talk* for pages. "Pages" is a literate term. This is a talk show. It's an oral form.

THOMAS: Well, someone must be writing it down. Otherwise it wouldn't be appearing here in this book.

HOST: Then I'll ask you this: Do you think the fact that our conversation is being written down changes its essential nature?

THOMAS: I doubt it.

HOST: Why?

THOMAS: Because I'm *doubting* Thomas, that's why. It's what I do—I *doubt.* It's part of my brand.

HOST: Wow. Your *brand*?

THOMAS: Yeah, things are going great! I even have an island named after me.

HOST: I know. Congratulations. Let's be serious for a minute. Here are a couple more things to consider. On the side of the Bible being a product of Orality:

The Bible contains no philosophical proofs about the existence of God, and no systematic theology. It is not always told in chronological order. Sometimes the same stories are told twice, but some of the details appear to vary. It makes no attempt to whitewash evil, nor the failures of its heroes, whether individual or collective. It describes miracles that seem preposterous to the modern scientific mind. It is told by different people from different nationalities from different eras, in multiple genres. Its chief aim appears to be relational. Rather than presenting an analytical argument, it tells the *story* about the God who created the universe wanting a loving relationship with us, and wanting us to love each other. If it were primarily a product of literacy, it just would not be set up this way.

THOMAS: And yet it still is a book.

HOST: Not only that, but it is the most read and widely analyzed book in history. Unfortunately, all this analysis has led to all kinds of conflicts, and ironically, moved us down the continuum away from God and each other, sometimes even to the point of war. It's not that there is anything wrong with analysis, per se, or that it cannot be beneficial. It's just that analysis is not the main point.

Interestingly, the Bible itself directly warns about our tendency to let analysis drown out its relational message. *There is no end to the writing of books, and too much study will wear you out,"* says Ecclesiastes. The Pharisees were condemned because they knew the scriptures backwards and forwards, yet missed their relational point completely. Jesus frequently quoted the scriptures, but his primary focus was on loving the people around him. Even Paul, the most literate of the Bible writers, says that even if we have knowledge, without love we are nothing.[17]

THOMAS: Then it seems like we have a paradox. The Bible is a product of both Orality and Literacy at the same time. It moves us down the continuum, but at the same time it brings us closer to God. How is this possible?

HOST: Once again, because it is unique. If you are willing to say that a piece of writing can be divinely inspired, you're talking about a form of language that has a different type of agency.

THOMAS: What do you mean?

HOST: That it can't be judged by any ordinary criteria. With any type of writing, the words travel from the mind of the

---

[17] I Corinthians 13

author onto the page, only to be recreated again in the mind of the reader. With scriptures, people of faith believe the Spirit intervenes in the process.

THOMAS: What if you don't believe that writing can be inspired? I've been known to have a hard time believing things sometimes.

HOST: Like everything else that is spiritual in nature, you have to be willing to listen. God does not force himself on us. Scriptures call for a different kind of reading and listening—a whole different kind of awareness. It's interesting to note that through the centuries, the more people have tried to look at the Bible solely through the analytical lens of Literacy, the more it has kept slipping out of their grasp to become relational in the hearts and minds of its readers. It seems that its inspired nature has continued to confound all efforts to constrain it.

In the words of *Lamentations*, it is new every morning.[18]

St. Thomas

[18] Lamentations 3:23

ADAGE

Just because it's new
Doesn't make it good.
Just because you can
Doesn't mean you should.

C.M. Collins

QUESTION FOUR

HOW DOES TECHNOLOGY AFFECT OUR FREEDOM?

We were all supposed to be working part-time by now. In a famous essay written in 1931, economist John Maynard Keynes predicted that due to technological advancement and the magic of compound interest, his grandchildren would someday live in "an age of leisure and abundance" where people would only have to work three hours per day.[19] Freed from the shackles of economic necessity, they would become paragons of virtue who would turn away from the love of money and care more about their fellow human beings instead. So, let me ask you—how's *your* fifteen-hour workweek going? And just where are all of these altruistic people?

Instead of being free, we are now more dependent on, co-dependent with, and addicted to our technological devices than Keynes could have ever imagined. Nor could he have imagined the hold that large-scale technological systems would have on the entire structure of our culture. As Neil Postman says, people have "come to love their oppression, and adore the technologies that undo their capacities to

---

[19] *Economic Possibilities for Our Grandchildren.*

think."[20] What happened? Why didn't Keynes's predictions come true? Let's take a look.

CONDITIONING

The dawning of the twentieth century was an important time in the history of technology. The Industrial Revolution, which by that time had pretty much run out of steam, met up with the newly developed principles of Mass Production and Behavioral Psychology, and things really began to take off again.

Up to that point, frugality had been seen as a virtue. Common proverbs taught people to *Waste not, want not,* and that *A penny saved is a penny earned.* "Consumption" was the word for pulmonary tuberculosis. In church, people learned that it was harder for a rich man to go to heaven than it was for a camel to go through the eye of a needle. It wasn't that people didn't appreciate wealth, but since it wasn't realistically possible for most of them to have it, they had learned to find their happiness elsewhere.

Mass production changed all this. For the first time in history, it was possible for the average person to have an overabundance of things that he or she didn't need. *Lots* of things. But the great industrialists of the era soon realized they had a problem: people had been taught to be frugal for so long that they had no use for all that extra stuff and really didn't want it. Something would have to be done to change their values.

---

[20] *Amusing Ourselves to Death.*

Enter Ivan Pavlov. The Russian scientist's famous experiments with dogs showed that by pairing the sight and smell of meat with something like the sound of a bell, a dog could later be manipulated into salivating simply upon hearing the bell.

It turns out that Pavlov's methods were even more effective on people than on dogs, because in addition to biological needs like hunger, people also have sophisticated emotional and social needs that can be manipulated too.

Few understood the implications of this idea better than Henry Ford, founder of the Ford Motor Company and creator of the Model-T. People knew they didn't need a Model-T, of course. History had gotten along just fine without them. So Ford, along with his head of advertising, Nevel Hawkins, decided to take a far subtler approach at convincing people to buy one.

A Ford Motor Company advertisement might picture a beautiful actress sitting on the fender of a Model-T. The presence of the woman would invoke an involuntary biological response in the typical male, who would want to be with her, and an involuntary social response in the typical female, who would want to be her. Either way, the beautiful actress would be paired with the car in their subconscious minds. Later, when a Model-T rumbled down the street next to them, the man would respond to the car biologically and the woman socially, leading both to want to buy one.

The great power of this approach was that people could be conditioned to want things that their rational minds told them they didn't need. Frugality, as a virtue, could be eliminated, without anybody realizing what had happened.

According to historian Steven Watts:

> "By the early 1900s advertisers had begun using brighter cultural colors to portray commercial goods as conveyors of emotional happiness, personal desire, and private satisfactions. Advertising increasingly appeared as a kind of commercial therapy that promised varieties of self-fulfillment: fantasies of play and fun, possibilities of romance, excursions into progress and modernity, pathways to increased social status."[21]

Did it work? Yes—in spectacular fashion. After the implementation of this type of advertising, sales of the Model-T multiplied 132 times, from 6,181 to 815,920 cars per year,[22] even though nobody still really needed one.

The Ford Motor Company was not alone in this venture. All across the corporate landscape, Pavlov's principles were rushed into play. Business boomed. It turns out people could be manipulated into buying just about anything. Advertising soon became an essential part of business, and eventually became an industry of its own. Ads sprouted up everywhere, like weeds. By 1985, Postman was able to make the startling observation that the television commercial was "the single most voluminous form of communication in our society."[23] Today, global advertising has become a $500 billion per year industry. It is estimated that the average young person is exposed to over two million advertisements by the age of eighteen.

The result has been a radical restructuring of societal values. The mass consumer culture is here to stay; the days of

---

[21] *The People's Tycoon: Henry Ford and the American Century.*
[22] Hawkins, Nevel A.
[23] Postman, ibid.

frugality are over. "Nothing could be more splendid than a world in which everybody has all that he wants," Ford proclaimed.[24] Technology was the tool he used to provide it.

Repetition after repetition after relentless repetition:

> "You are unhappy. There is something wrong with you. You are not (*smart, cool, sexy, healthy, rich, popular, athletic, happy, successful*) enough. But never fear, Product X is here. See all the shiny happy people in the ad? They *are* (*smart, cool, sexy, healthy, rich, popular, athletic, happy, successful*). You will be like them if you buy Product X too. Do not stop to think about how absurd this is."

The basic goal of all advertising is to keep us in a perpetual state of dissatisfaction about ourselves and our lives and to get us to believe that the only way out is by buying things. This message gets pounded into us from the day we are born until the day we die. No other people in the history of humanity have ever had to fight off such a powerful and constant wave of propaganda.

Since advertising's effects are largely subconscious, we are often unaware that we are being manipulated. According to Hawkins, the goal for a salesman at Ford was to present a product as a suggestion that "gets in unawares and makes itself at home in the mind of the prospect so effectively that he believes it is one of his own family of ideas."[25] This is frightening. Consider what happens to the child who hears

---

[24] Watts, ibid.

[25] Hawkins. ibid.

there is something wrong with you two million times before the age of eighteen, and then thinks it was his own idea! The effect is made even worse when adults around the child start celebrating that same message. Witness the rather bizarre spectacle of the hype surrounding Super Bowl commercials each year, where adulation is poured out, like saliva from Pavlov's dogs, on one of the main sources of our children's depression.

Things got even worse when the internet came along. Because advertising bypasses the rational mind, the internet is the perfect place for it. Its infinite nature can host an infinite number of potential ads. Its non-linear, image-based environment dovetails seamlessly with the irrational types of manipulations advertisers love to use. And since it democratizes all information, the ads themselves often become as important as the intended content. Advertising has never had a more hospitable home.

Now that there are more and more ads on the internet, therefore, we can expect that advertising will become even more irrational. We are already starting to see this. The beautiful actress sitting demurely on the fender of a Model-T has been replaced today by a scantily clad bikini model suggestively fondling a hamburger. Other ads have gone completely non-linear: in an ad for a car company, a moody actor drives through the desert night muttering to himself. Why? It's not exactly clear. The next step in advertising will come in the form of what Aldous Huxley called the *feelies*, where electrodes and sensors hooked up to the body will allow viewers to physically experience all the events appearing on their screens. Many people will never want to leave the advertising environment.

How does all of this affect our freedom?

Let's go back to the farmer from the Middle Ages that we mentioned earlier. Imagine how quiet his life must have been. Imagine the lack of visual clutter. Imagine having the uninterrupted time to think. Imagine being able to make decisions without the relentless drumbeat of irrationality and materialism that constantly pounds at the back of our minds today: One million. Two million. Three million. Five million. Ten million, million repetitions, each one telling us that there is something wrong with us, that our salvation can only come through the purchase of products and worship of gods we know to be false. Are we happy? Are we free? The question is absurd. We're *suffocating*. It's no wonder Keynes's predictions never came true. The farmer was the one who was free—not us.

DEPENDENCE

When we first bought our house it was the worst one on the block. The previous owner was an older woman who had basically let the place go when her husband died, and it was a mess. Every drawer was stuffed full of trash, piles of dirty laundry lay everywhere and the walls were stained where the roof had been leaking. There was a pool in the backyard, but it was covered by a mysterious greenish growth two inches thick so you couldn't see to the bottom.

For us, it was perfect. We were young, pretty handy with a paintbrush, and the elementary school for the kids was just down the street. Little did we know that repairing and remodeling a complete fixer-upper is a Sisyphean task—just when you've pushed the boulder all the way to the top of the hill, it rolls back down and you have to start all over again. Nevertheless, painstakingly, over the course of many, many years the house has taken shape. The last room to be renovated was the studio. Since it involved a special ceiling that required a lot of screws, I finally relented and bought a battery-powered screwdriver.

What a revelation! What had I been thinking all those years, screwing everything in by hand? The ceiling installation went smoothly and relatively easily, and in the process I became a convert to the joy of small power tools. Until a couple of weeks later when I had to do a tiny repair around the house. I grabbed an ordinary screwdriver to do a job I had done many times before. This time, however, I had a hard time. The screw wouldn't untighten going one way, and screwing it back in the other way took a lot of effort. My hand got tired and sore. I thought about getting the battery-powered screwdriver out, but it was in the garage, and besides, I shouldn't have had to use it to do a job so small.

But time had passed and the paradigm had shifted. I gave in. I went out to the garage, got the new screwdriver and finished the job quickly, but sheepishly. I had arrived at the point where even if I wanted to go back and use the old one, it simply required too much effort. What I had gained in efficiency and ease I had lost in hand and wrist strength. In a shockingly short amount of time, I had become dependent on a new technology.

The work itself had not changed. A screwdriver was still a screwdriver and the screw still needed to be screwed in. What had changed was my perception. What had been easy before now seemed difficult; I had become soft. People were not always like this. Laura Ingalls Wilder talks about her Pa walking three hundred miles across the Great Plains to find a job. We read about Civil War nurses amputating soldiers' limbs on the battlefield using no anesthesia. If you shook the hand of a blacksmith from 1750 you'd have to be careful that he didn't break all the bones in yours. I don't think any of them would have had trouble with my little job.

And their strength was not just physical. They understood that life was difficult and full of hardship because they faced it daily. As a result, they had a courage and resolve that we seem to have lost today, where a short car ride without air conditioning throws us for a loop. It may be a cliché, but it really is true that hardship produces character. Eliminate all of life's hardships, and there's very little of it left.

But life *is* suffering. It is the first of the Four Noble Truths that Buddha taught. It is only by going through suffering that we are released from its power over us. This is one of life's most important spiritual lessons. The apostle Paul talks about welcoming suffering, "because we know that suffering produces perseverance; perseverance, character; character, hope."[26]

"I will free you from the difficulties of the task at hand," says technology, "and this will make you temporarily happy, for your life will be easier. But there will be a price to pay. At each step, I will incrementally rob you of your strength, your character and your freedom. In your weakened state, you will be less able to handle the next difficulty that comes around. Having sought me out before, you will turn to me again, and again, and then again. In the end, you will be utterly dependent and unwilling to function without me."

Life itself has not changed. It is still difficult. But it is still meant to be lived freely. This isn't just about screwdrivers, of course—it's about how we give up our freedoms every time we tether ourselves to technology. The more powerful and

---

[26] Romans 5:3-4

invasive the technology, the weaker we become. Call it *dependency in the name of expediency.* We should proceed with caution.

CODEPENDENCE

One of the best ways to describe people's relationship with their devices today is *codependent.* It is a curious sort of codependency, however, since the devices are inanimate objects. Theoretically, this shouldn't even be possible. When we remember that technology is a movement that has implications for relationships, however, it's really not as strange as it first sounds.

For the purposes of this discussion, I have taken a description of codependency and adjusted it for the technological age, substituting *technological device* for *person:*

> "Codependency is a type of dysfunctional helping relationship where a technological device supports or enables a person's immaturity, irresponsibility, or under achievement. The most common theme is an excessive reliance on the technological device for approval and a sense of identity. A co-dependent person is someone who cannot function from their innate self and whose thinking and behavior is instead organized around the technological device."[27]

Are you and your device codependent? Here's a short quiz to help you find out:

---

[27] Lancer, Darlene; Johnson, Skip R.

1. Do you feel compelled to look every time your device summons you?
2. Do you constantly seek out your device when you are bored?
3. Do you trust your device's opinion on things more than your own?
4. Do you ever use your device to avoid conflict or uncomfortable situations with others?
5. Are you overly preoccupied with your device's well-being?
6. If you are away from your device for any extended period of time, do you experience feelings of anxiety? Abandonment?
7. Do you get self-esteem and a sense of identity from having purchased your device?
8. Do you get jealous or feel neglected when you see your device in someone else's hands?
9. Do you feel uncomfortable interacting with others unless your device is also present?
10. Have you ever experienced romantic or sexual feelings toward your device?

To boil it down to one simple question: Is your device the first thing you think of when you wake up in the morning, and the last thing you think of before you go to sleep at night? If so, you may be codependent.

Simplistic? Perhaps. The danger with quizzes such as these is that they tend to stay at the surface level and never get to the underlying causes of the problem. They also cannot account for unique pathologies that may arise in individual cases. Nevertheless, your answers were quite enlightening, weren't they?

At first, codependency was seen solely as a woman's issue. It focused primarily on wives who were emotionally reliant upon their alcoholic, abusive husbands. The relationships were *co*-dependent because the husbands' pathologies were dependent on the enabling attitudes of their wives. Technology has now thrown the codependent floodgates wide open. Since the devices are inanimate objects, *all* people, regardless of gender, ethnicity, religion, socio-economic status, age or sexual orientation can be in a relationship with one. Let's all say it together: *I'm technologically codependent, you're technologically codependent, we're all technologically codependent!*

This is pretty twisted and weird. How is it that we are so devoted to these little boxes of metal and plastic? The answer is that we don't think of them as such. One of the most troubling aspects of humanity's interaction with machines is our tendency to attribute human characteristics to them. To us, our devices are not just little boxes. Instead, they are people too.

This was first discovered early in the development of artificial intelligence (AI). In 1966, computer scientist Joseph Weizenbaum developed a computer program called ELIZA.[28] One of its scripts was named DOCTOR, and was designed to mimic the typical responses of a Rogerian psychologist. The "patient" at the keyboard would type in "I feel depressed" and the program would answer something like "I am sorry to hear that you are depressed. How do you think your coming here will help?" "Maybe it will help me get along better with my mother," the patient might respond. "Tell me about your mother," the computer would answer, and the conversation would be off and running. His whole intention in creating

[28] *Computer Power and Human Reason.* This is the single best book I have read on the relationship between people and computers.

ELIZA in the first place was to show how shallow human-computer interactions really were.

He didn't anticipate the role people's emotions would play. Supporters of Rogerian therapy say that it is the empathetic environment created by the therapist that makes the treatment so effective. When the first trial users of ELIZA started reporting positive empathetic results, Weizenbaum was astonished. How could a computer program, that could not *by definition* show any empathy, be achieving this?

The answer was that almost from the first moment they sat down and started typing, users began attributing a whole host of human characteristics to ELIZA, including empathy and intelligence. They became attached to "speaking" to it. Weizenbaum was startled one day to come upon his secretary deep in dialogue with the machine. When he walked in the room, she quickly shooed him away, saying that the conversation had become "private".

In his book, Weizenbaum says,

> "What I had not realized is that extremely short exposures to a relatively simple computer program could induce powerful delusional thinking in quite normal people." [29]

Surely his secretary knew she was talking to a computer. After all, she had witnessed ELIZA's day-by-day development. But her ability to reason was overwhelmed by her emotional needs, which she somehow convinced herself were being met. The trouble is, they weren't. She was really just becoming

[29] Ibid.

codependent. Just like the wives of the abusive husbands, she was attributing empathy to an essentially un-empathetic source. The more Weizenbaum's secretary sought out the computer, the more she became convinced it was a person who cared. The computer, of course, was dependent on her in order to get turned on. [30]

Remember too that this happened back when computers were as big as a room. Now that everything from cars to personal digital assistants come with a kind, understanding, and even sexy vocal interface, the emotional lines between humans and machines are increasingly becoming blurred. AI expert David Levy predicts that by 2050, falling in love with, marrying, and having sex with robots will be commonplace, and will be regarded as "normal extensions of our feelings of love and sexual desire for other humans."[31]

Does this sound disturbing to you? Then this may be the point where you will feel the need to opt out. For many people, however, the step will be a small one. If you've already had an intense, emotional involvement with your device, how great will it be to finally feel its gentle, loving touch?

But true communion is always reciprocal. A robot "spouse" will never be able to love us back, no matter how closely it has been programmed to ape human actions. The fact that people will *believe* their emotional needs are being met will just place them, by definition, all the more squarely in co-dependent territory.

The choice is up to us. The second we stop needing our devices, we free ourselves from the type of delusional

---

[30] The co-opting of words can interestingly go both ways.

[31] *Love and Sex With Robots: The Evolution of Human-Robot Relationships.*

thinking Weizenbaum describes, and from codependency. If we are not able to break away, we will continue to move down the line from dependency to codependency and finally on to full-blown addiction.

ADDICTION

The next step toward the elimination of our freedom is technological addiction. In the past, people were hesitant to use the term. The pathology was new and was hard to quantify. People were understandably reluctant to describe their relationship with their devices as a problem in the same way they would alcoholism, drug addiction, or compulsive gambling. Certainly the tech industry was not (and is still not) interested in having their products labeled as addictive, lest they be held accountable for them, like Big Tobacco. Even the health industry was hesitant to come on board because they knew they would have an epidemic on their hands. But the facts are now becoming clear. Technological addiction is real, is increasing, and is rapidly becoming a major societal issue.[32]

As with all other addictions, tech addiction is hard to precisely define and measure. At what exact point does tech dependence turn into tech addiction? At what exact point does recreational drinking become alcoholism? There is no clearly defined line for either, only a set of variables to be interpreted. The single aspect that all addicts share, however,

---

[32] For excellent overviews of the current research, see *i-Minds*, and Brand, Matthias, et al.

is a lack of control. The rewards of engaging in the addictive activity are so intense that a person can no longer disengage from it without tremendous effort, and often only with the help of others. The activity or substance, not the person, is running the show.

Unfortunately, addictions are easy to hide, often until the damage has already been done. This is especially true with tech addictions, since sitting in front of screens now has such high societal approval. Other factors, like a lack of self-awareness or having a weak support system often mean that an addict will not seek the treatment that he or she needs. Sometimes a person keeps an addiction hidden out of shame.

The first authoritative test for internet addiction (the IAT) was developed in 2011 by Dr. Kimberly Young, and is similar to the types of tests given for other addictions. It has proven to be very effective at diagnosing the problem, but it does have one significant drawback. Like the other tests, many of its questions relate to how a subject's personal relationships have been affected by the addiction. But what if a person has no personal relationships to begin with? This is such a big problem with tech addicts that psychologists must take this factor into account in order not to skew their test results.

This is where brain studies can be helpful. The front part of the brain, or prefrontal cortex, is the center of executive control—the place where we make rational decisions and control our impulses. As an addiction sets in, the prefrontal cortex progressively loses its ability to regulate behavior. The dopamine reward system is also altered, leading a person to feel good only when participating in the addictive activity. This leads to a physical "rewiring" of the brain. [33] A

[33] Another tech word co-opted to describe a natural phenomenon.

neuroscientist can look at a brain scan and see the physical structures that have changed due to an addiction. In this regard, the brains of tech addicts are remarkably similar to the affected brains of all the other types of addicts.

But not all tech addicts are created equal. Although everyone is susceptible to the addiction, some are more susceptible than others. Children appear to be the most at risk, since their prefrontal cortexes are not yet fully developed. Others may have a genetic predisposition for tech addiction, but the studies are not yet conclusive. For many people, certain cultural factors can also come into play. It turns out that tech addiction is not the exclusive province of the stereotypical lonely male. Anyone who feels alienated from society or is suffering from depression or loneliness is more likely than others to turn to technology as an escape. Studies have specifically identified recent retirees, newly separated or divorced females, and LGBT teens from small communities as being more at risk for tech addiction, for example.[34]

There are also different types of tech-influenced addicts. The first are the kinds of people who would have been addicted to things like alcohol, drugs, or gambling anyway, with or without the technological help. The internet just provides them with easier access to the addictive substances or experiences they crave.

The second group is made up of people who would not be addicts at all had technology not provided them with the opportunity. There are many people who would not go into a strip club, for example, who are addicted to internet pornography in the anonymity of their own homes. Online gambling among college students often fits into this category.

---

[34] *i-Minds.*

Then there are the people for whom the technology itself has become addictive. The most well known are video gaming addicts, about whom much has been written elsewhere. It turns out, however, that *anyone* who spends an inordinate amount of time online doing *anything* is at risk. This is because internet searches themselves are addictive. They operate on the same variable reward system as slot machines in casinos. When a person plays a slot machine, there is a moment of anticipation as the wheels spin and the matches are revealed. A chemical jolt floods the brain, creating what in casinos is called the "gambler's high". The feeling is so intense that people seek it again and again. It happens with internet searches as well. People can't explain why they feel so compelled to check and re-check their e-mail, their bid on eBay, or whether or not their social media posts got any "likes". Teachers report that certain parents check their child's on-line grades obsessively throughout the day—every single day. Other people will sit for hours randomly clicking from site to site with no purpose at all in mind. None of this makes any sense until it is seen in the light of chemical addiction.

Finally, the fourth and saddest group is made up of those who are addicted to both the substance or activity *and* the technology that delivers it. They are the ones who are the hardest to treat.

Just how many tech addicts are there? As with any addiction, it is difficult to say precisely. The problem is made harder because we, as a culture, simply don't *want* to believe that these devices which are our constant companions, forsaking all others until death do us part, can actually be bad for us. In this, we are lagging behind the curve. Australia, China, France, Germany, India, Italy, Japan, Taiwan, Singapore and South Korea have all identified technological addiction as a major health issue, and are opening centers and developing treatments to deal with it.

The great technological robber barons of our era would prefer that you not think about any of this. They *really* don't want the schools to wake up about tech addiction, since the corporate strategy to target the education market has been so effective. It's ironic that the streets around our schools are set apart as drug-free zones, yet within those confines the children are forced to sit all day in front of addictive screens. By the time the students raised in our schools are in college, the average male spends eight hours per day on his phone, and the average female ten.[35] The school was often the dealer that gave them their first free sample.

To find out whether or not you are a tech addict you can try the technique used by Dr. Young in her treatment centers. The first thing required of patients is that they go on an immediate and complete 72-hour "digital detox". No phones. No computers. No TV. Nada. You want to claim that you have no problem with technological addiction, and that you, not your device, are in charge of your life? Great! You've been challenged. Try the detox. If you find yourself becoming irritable or anxious, or suffering from symptoms of withdrawal, guess what?

---

[35] Roberts, James A. et al.

iOS iWork iLife iCloud iMac iBook
iPad iPod iPhone iWatch iOS iWork
iLife iCloud iMac iBook iPad iPod
iPhone iCloud iWatch iOS iWork iLife
iCloud iMac iBook iBook iPad iPod
iPhone iCloud iWatch iOS iWork iLife
iCloud iMac iBook iPad iPod iPhone
iWatch iOS iWork iLife iCloud iMac
iBook iPad iPod iPhone iWatch iOS
iWork iLife iCloud iMac iBook iPad
iPod iPhone iWatch iOS iWork iLife
iCloud iMac iBook iPad iPod iPhone
iWatch iOS iWork iMac iLife iCloud
iWork iPod **iSolation** iCloud iMac
iBook iPad iWatch iLife iOS iPhone
iOS iWork iLife iCloud iMac iBook
iPad iPod iPhone iWatch iOS iWork
iLife iWork iMac iOs iBook iPad iPod
iPhone iWatch iPad iOS iWork iLife
iCloud iMac iBook iPad iPod iPhone
iWatch iOS iWork iLife iCloud iMac
iBook iPad iPod iPhone iWatch iOS
iWork iLife iCloud iMac iBook iPad
iPhone iLife iOS iWork iLife iCloud
iMac iBook iPad iPod iWatch IPhone
iBook iPad iPhone iOS iPod iWatch

SYSTEMS

Technology is most effective at robbing us of our freedom when we forget that it is there. Ironically, the bigger a technology gets, the more it becomes invisible. As it grows, it integrates itself so effectively into the world that we have a hard time perceiving of it as anything but natural. This is when it is at its most dangerous.

Consider, for example, Lewis Mumford's famous observation about the mechanical clock. Developed by monks in the Middle Ages to regulate the intervals between their prayers, the mechanical clock eventually became "the key machine of the modern industrial age."[36] Virtually all technological innovation since then rests upon our ability to precisely measure time. Without the clock, there would be no internal combustion engine, no microwave oven, no computer games, and no thermonuclear warfare.

Keep in mind that we can't even really explain what time *is*. Nevertheless, we have given it control over virtually every aspect of our lives: we measure it, save it, kill it, waste it, and watch it fly when we're having fun. It has become a virtue to use it wisely, and you'll be marked tardy to first period if you're not in your seat by the time the bell rings. We complain constantly about not having enough of it, yet seldom question why we ever chose to measure and confine it in the first place.

And "chose" is the key word. Time used to be tied to nature. The day would start about when the sun came up and end about when it went down. People worked hard, but life was fine as long as the crops were harvested sometime before

---

[36] *Technics and Civilization.*

the snows came. They were not rushing around having to justify every split second of their lives. Now that our existence is regulated by the clock, we'd better be able to account for our every hour, minute, and second, and God help us if we can't keep up with the pace. Yet how often do we stop to think about how strange this is?

Sadly, all this came from a genuinely spiritual aim. The monks thought they'd be certain to be closer to God by developing the right tool to regulate their prayers. Instead, they exhibited one of the fallacies of everyone who tries to channel the Spirit though technology: every time we try to control the Spirit, the process ends up confining us instead.

This is because all technologies, if allowed to develop long enough, eventually become organic systems.[37] Because they grow out of their own inner dynamics, they quickly become too big to control and we, ironically, become subject to them. We've already seen how one simple rule can expand into an intricate set of laws. The same principle is at work with technology.

A SHORT PARABLE

A certain young man set out to create the world's greatest computer program. It was going to be the most powerful, the most efficient and the most useful that the world had ever seen. His name was Programmer A. Filled with a sense of great purpose and even greater expectations, Programmer A began his work. He spent his days, then his weeks and his months in front of his computer screen, writing and re-writing each line of code until it was perfect.

The harder he worked, the more ideas he had. The more

---

[37] For the best discussion of this, see Jaques Ellul's *The Technological Society.*

ideas he had, the more he wrote. The more he wrote, the bigger the program got. The bigger the program got, the harder he worked. Soon he was not only exhausted, but realized he was losing control of the variables and needed some help.

So he called his best friend, Programmer B. He asked her if she would help him with a certain section of the program that was giving him trouble. She said she would. As she began working, however, she soon began experiencing the same phenomenon A had. There seemed to be no end to the various possibilities she could explore. After getting A's permission, therefore, she went ahead and hired C. Meanwhile A had also hired D, E and F, who in turn hired G, H and I, and so forth. Soon the office was filled with an entire alphabet of programmers, each in charge of separate tasks.

Eventually the program reached a point of such complexity that no one person could fully explain what it did. A had no idea what Q was doing, and R had never even met Z. This complexity created instability. The program developed bugs. Hoping to capitalize on the work they had done, they released version 1.0 anyway, hoping that the public would help them fix it. The complaints poured in, but so did some of the solutions. Rather than going back to the root causes of each problem, however, they patched the program up externally. It grew and grew further still. Version 1.1 was released, followed by 1.2, 1.3, 1.4. . . and 2.0. Everybody knew there were still problems, but no one wanted to start over completely, and besides, each version was incrementally better than the last.

As the public began to use the program, they came up with other uses for it that the alphabet team had never even imagined, especially when it was used in conjunction with still other programs. Things grew exponentially even larger, with still more unpredictable results. Driven on by this internal logic, what started out as one person's single idea became an *organism*, ever growing, ever more powerful, ever more complex, even though no one was in charge of it.

And whatever became of Programmer A? After a few years he cashed in all his stock options and retired, and is now living somewhere on a beach in Mexico, regaling bemused locals with tales of how he changed the world.

Unrestricted technological growth often gets rationalized away. Since no one is individually responsible for the way a technology develops, no one person can be blamed if it goes awry. "It's not my fault things went wrong," a programmer will say. "All I did was sit in a room doing math problems. It's not my job to be a moral philosopher." In any large system the individual can disappear into a fog of moral ambiguity, like the soldier who justifies an atrocity because he was just obeying orders. This stance runs rampant among techies, since the capacity to deny culpability is a necessary prerequisite for many, many tech jobs. The arrow continues down the line, even though no one person is in charge of the movement.

This was what Ted Kaczynski, the Unabomber, didn't understand, perhaps because he was mentally ill. He thought that by killing individual people, he would open society's eyes to the dangers of the technological system, and that we would make changes accordingly. But the organism is bigger than any one individual. Even if Kaczynski had completely nuked all of Silicon Valley down to the ashes, technology would have just kept right on growing in Seattle, North Carolina, China, India and . . .

This is one key reason why anyone who thinks he or she can control technological outcomes is naive. As sociologist and author Jacques Ellul says:

> "The man who thinks he can choose between the good and bad effects of technology proves by this very idea that he is totally set within technical determination and that he is not yet even seeing the first step of freedom."[38]

This is a prescient warning to those who think that introducing technology into a place like a church or temple will have only the desired results. At a local church near us, the pastor introduced an internal e-mail system, hoping that the congregation would use it to share prayer requests, personal concerns and the like. Of course they did this. The unintended consequences, however, were that the system became a forum for gossip, and the people slowly stopped talking to each other face to face, since they had already communicated via e-mail. One can argue that this was the pastor's fault, but was it solely his? Or was it also everyone else's? Or was it simply the way things go? Whatever the case may be, what began with a spiritual aim quickly grew out of control, and ended up separating the people even further.

The more that various decisions are involved in the growth of a system, the more chance there is that the system will fail due to human error. The easiest way to avoid this, we are now being told, is to let the computers take control of the systems. Since they operate on perfect efficiency, logic, and

---

[38] *The Ethics of Freedom.*

reason, the thinking goes, they will be much better at handling tasks than a person ever could.

The trouble is, this only works for certain kinds of tasks. Computers can only "think" in mechanical ways. Human problems must therefore be converted into mechanical forms before the machines can solve them. But things like our spirituality cannot be reduced to a series of ones and zeroes. As we have said, no machine can account for the breath of God that is in us.

Because technology is a jealous god, it responds by striving to reduce all human thinking to only those parts that are purely logical and efficient, and then claiming that these make up all human thought. It is the ordination of insufficiency all over again. Since the computer can't experience God, it doesn't want you to either—but it will offer you a sparkly, shiny counterfeit of the real thing.

But mechanical thinking does *not* constitute all human thought. Efficiency is *not* inherently good for us. Sometimes life is messy. Sometimes our emotions rightfully hold sway. Sometimes it's better to do things slowly, or the hard way. Meaning itself cannot be reduced to mere function. Might does not always make right. We are children of the earth, but we are also the children of God.

Another element of spirituality that the computers cannot account for is freedom. As Weizenbaum says:

> "The statement that justice and freedom are better in themselves than injustice and oppression is scientifically unverifiable and useless. Endorsed by venerable documents, they may still enjoy a certain

prestige (but) they lack any confirmation by reason in its modern sense."[39]

In other words, if you're going to demand perfect efficiency, reason and logic, you're going to have to sacrifice ideas like freedom. Freedom is a belief, an assertion, a proclamation, a leap of faith. It is not self-evident. The computers can never experience freedom because they can only do what they have been programmed to do. Once we have given them total control, they will no longer allow us to either.

This is one of the reasons people are so deeply troubled by artificial intelligence. Any system working primarily on reason and efficiency, which sees all problems, and thus all solutions in purely mechanistic terms, *will* eliminate any extraneous human ideas—including freedom. Sure, your self-driving car will free you from having to drive, for example, but you will be totally dependent on its functions to get to your destination. If it "decides" to take you somewhere else, it will. The freedom it offers is an illusion.

And what happens when the machines make decisions that are immoral, or wrong, or dangerous? Words like "immoral", "wrong", and "dangerous" are also strictly abstract human terms that have no coherency in the mechanistic scheme of things. And since no one will completely know how the machines work, no one will know how to stop them.

This is one of the reasons why technology *must* co-opt the language. Terms from nature, humanity and spirituality must be redefined in order to hide the larger technological agenda. As we move further down the line of the continuum, the

---

[39] Weizenbaum, ibid.

systems must be kept invisible.

Technology may not be spiritual, but it sure can be religious. Or maybe it's the other way around. Maybe religion is a technology. They both seem to work in mysterious ways. Ellul says that when tech systems grow large enough, they take on a sacred nature.[40] Hmm . . . let's see. Almighty powers that no one fully understands. Sovereign entities that control our daily lives. The illusion of order in a world of chaos. The promise of a glorious future to come. No wonder the people worship and bow down.

Technology, not religion, has become the new opiate of the masses. [41]

Is it even possible for a person living in our technological age to be free? Technology has conditioned us to believe things we know are not true and buy things we know we don't need. It has made us physically dependent, emotionally codependent, and has turned us all into addicts. Even when our intentions are good, every attempt we make at harnessing its power blooms into huge systems that end up imprisoning us even further. Its philosophies fail to console us, so we end up worshipping either ourselves or the idols of our own making.

---

40 Ellul, ibid.

41 A quick, perhaps unrelated thought: I wonder if the current preoccupation with trying to prove that all of our choices are biologically determined is really more of a matter of *Credo Quia Consolans* (*I believe it because it consoles me*), rather than a dispassionate search for truth. After all, if I have chosen to relinquish my freedoms to the machines, wouldn't it be nice to have some biological excuse for having done so?

The systems are everywhere:

> The public education system, the penal system, the welfare system, the health care system, the political party system, the economic system, the interstate highway system, the postal system, global positioning systems, the NSA, Big Oil, Big Food, Big Agra, Big Pharma, the tax code, the advertising industry, the left-wing media, the alt-right media, social media, various religious denominations, and the societal expectations of your town.

We are trapped by all of them, like bug specimens pinned to a card. Technology is intimately intertwined with every single item on this list, yet it keeps on promising that it is going to set us free. This makes absolutely no sense at all, yet we keep on believing it.

You are right if all of this feels suffocating. It is. Thankfully, all is not lost. We are trapped on the continuum, but there is still hope for us. That hope cannot, however, rest in technology.

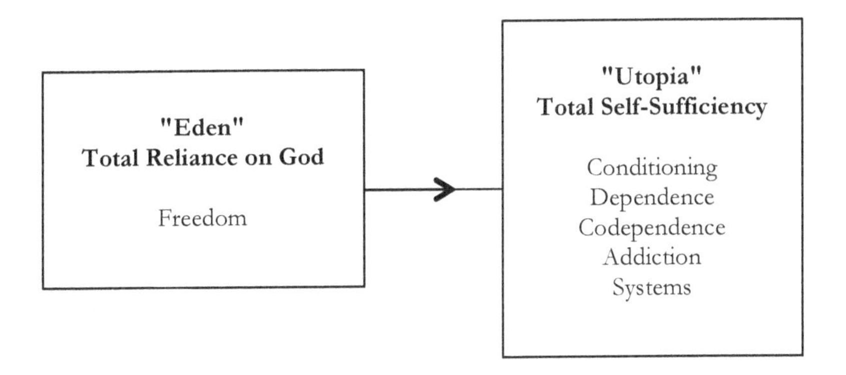

WANG CH'UNG

HOST: Welcome back, everybody. Let's go on to our next caller. It looks like we're in for another interesting conversation. On the line with us now is first century Chinese philosopher and skeptic Wang Ch'ung. Welcome to the show!

WANG: Don't start.

HOST: Don't start what?

WANG: Look—everything in my life was going along just fine—right up until 1986. Since then, every time I meet people of a certain age, the first thing they ask me is if everybody is going to have fun tonight. It gets old really quickly.

HOST: As a child of the 80s, I know exactly what you're talking about. I won't ask, I promise.

WANG: Thanks. I get tired of being bombarded with pop cultural references all the time. Don't get me wrong—the band is great. But they haven't been around as long as I have!

HOST: I'm just excited to have you on the show. Do you have a question for me?

WANG: I do. I'd like to go back to a point you made earlier. You said that while an online ministry might be able to do some good in certain very limited circumstances, meeting together face-to-face was much better.

HOST: Right.

WANG: Well, I have a pastor friend who claims that the very most effective ministry he's ever had has been with members of the online gaming community he is a part of. He says that people are opening up to him like never before, and that he's been able to have the most authentic spiritual conversations with people that he has ever had. What do you think?

HOST: I find it to be profoundly sad.

WANG: Why?

HOST: The most effective ministry he's ever had is with people that he's never even really met? People who could be faking their entire personas? There's no way for him to know if any of it is even real. And that's been his *most* effective ministry?

WANG: I guess so. I'm just telling you what he says.

HOST: Here's the problem, and it's a common one. I hear this line of questioning all the time: "But what about technology X?" (In this case it's your friend's online gaming "ministry").

Then the person starts touting all of the good things that X can do. The follow-up question is always, "Are you suggesting we ought to get rid of X, even though it does all of this good?"

WANG: What's wrong with that?

HOST: It is the wrong question to ask. Of course the individual technology does some good. As long as people keep their focus on the immediate effects in front of them, however, they will never even think about the systems behind them—the systems that will end up enslaving them in the end.

WANG: Can you explain that a little further?

HOST: Sure. It's one thing to look at an app on your phone and say, "Wow! This app makes little angry birds fly around! Isn't that great?" The phone *is* going to do whatever it is supposed to do, in the short run. But this is only a positive if you view the game in isolation. What if before every use of their phones, people would say, "I have been irrationally conditioned to love this machine. I may be in a co-dependent relationship with it. I am probably addicted to it. Certainly it is habituating me to the artificial tech world, and taking me away from the real one. This machine is robbing me of my freedom, and altering my ability to think clearly. I should proceed with caution." Do you think it might take some of the fun out of playing the game?

WANG: People are never going to do that. They look at their phones over a hundred times a day.

HOST: That's my point. In a very short span of time, the entire culture has given their lives over to these machines, yet very few people are stopping to think about the larger impact this is having on them as individuals or on us as a society. It is unprecedented in human history. Never before have so many people been so firmly tethered to devices that are so disruptive to every aspect of their lives.

WANG: But people don't believe they are dependent or addicted. They claim their phones are giving them more freedom.

HOST: Sorry, but that's just not true. Now that the studies are catching up to the technology, it is no longer possible to hide behind that kind of statement with any degree of integrity. Besides, we don't even need the studies to show us this. Look all around you. People walk around with their noses constantly in their phones, often with headphones blasting as well. The entire culture seems intent on *not* engaging with the world. Or each other.

WANG: Do you think this involves some kind of hatred of the world? Or is it fear?

HOST: It's both. Because people aren't looking up at the world around them anymore, they're starting to develop irrational fears about it. This then drives them further into technological isolation, looking for safety. "No, Little Johnny, you can't go out and play in the street. The real world is a big, dangerous place. Come inside and play video games instead, where I can watch over you and protect you." Remember, technology grows like an organism, by its own interior logic.

Little Johnny learns to fear the world, and those things we fear, we often learn to hate. No wonder Little Johnny keeps his headphones on all the time.

WANG: So what about my pastor friend?

HOST: The first thing I'd do is talk with him about technological dependence and codependence. Then I'd give him the test for internet addiction. It may be that he is addicted to his online gaming world, and is trying to seek some kind of justification for having given in to the seduction. Next, I'd have him read all about catfishing and have him talk with people whose lives have been ruined by scammers. From there, we'd talk about the ways that words like "conversation", "authentic", "open" and "spiritual" have been co-opted.

As we delved even deeper, we'd talk about how every time he thinks he is helping someone online, he is also conditioning that person to seek the man-made online world instead of the Spirit. "I will lift mine eyes to the screens, whence cometh my help," the person will learn to say, and will turn to the screens again and again for spiritual guidance. This places your friend—a pastor no less—in the very awkward position of helping someone increase his or her faith in the works of man, rather than in God. The sad part is, the more "effective" your friend is, the worse he makes it.

Finally, I would say this: Being a pastor is a tough job. You are in a group of one, always having to be a moral and spiritual example for everyone around you. You have no one to really open up to about your own struggles, so you long for a place where you can anonymously say the things you really feel, without always having to worry about making someone mad or causing them to stumble spiritually. It is no surprise that the lure of internet temptations is especially hard for pastors to resist. I get it. However . . . the authentic spiritual truths, the ones you desire most in your heart of hearts, are always better than the counterfeits offered by technology. The face-to-face relationships you can have with real people, the ones who really love you, can be so much deeper than any relationships mediated by machines. So don't be afraid. Be willing to be vulnerable to the people around you. Don't *settle*. Step out in a leap of faith and your efforts will not be in vain. It's just like in Narnia. The counterfeit Turkish Delight you thought was so delicious was really making you sick all the time.

WANG: I love those Narnia stories.

HOST: You do? I guess I'm going to have to stop being surprised by anything you know.

WANG: Guess what else? These are Dance Hall Days, love. Dance all day long! See ya! I gotta book it!

HOST: Hey! Whatever happened to no more 80s?

Host's Note: Unfortunately, no images of Wang Ch'ung have survived from antiquity. Historians tend to credit him, however, for developing early prototypes of both the mullet haircut and acid-washed jeans.

TECH : RELIGION : TECH

BURNING MAN

Technology and religion work in very much the same way—they both stem from our desire for self-sufficient control, and they both grow by their own inner resources to become all-enveloping systems. While we've seen this over and over again lately with technology, it's far more rare to be able to watch first hand as a religious movement grows from an idea into a system. The last few decades have afforded us just this opportunity, however, in the form of Burning Man, one of the most prominent new religious movements of our time. What began on a San Francisco beach one evening in 1986 as a "playful impulse", in the words of its founder, Larry Harvey, has grown into a bona fide organized religion.

At first, its founders weren't even sure why they did it. Operating strictly on impulse, Harvey and friend Jerry James built an eight-foot high wooden statue of a man, drove it to a beach in San Francisco, doused it with gasoline, and set it on fire there on the sand. The result was unexpectedly profound. As they and the other people who happened to be on the beach that night gathered around the flames, they all felt that something significant was taking place. One man spontaneously broke out into song. A woman tried to walk up and hold the burning man's hand. They all left that night feeling that something inexplicable and mysterious had occurred.[42]

Had they left well enough alone, the moment might have stood as a genuinely serendipitous spiritual experience. But that's not what we humans do. Having captured lightning in a bottle once, the founders just had to try to recapture it again the next year. And the next. And the next. Each year the wooden man got bigger and the crowds grew larger. As more and more people started coming, the event began to spiral out of control. The gathering got rowdier. People of questionable intent began showing up. The cops got nervous. Permits were mandated. There began to be a need for *rules.*

To get out from under the city's restrictions, the event moved out to the Nevada desert. The multitude grew from hundreds to thousands. Soon, a list of Ten Commandments (called "principles") was produced. The organizers began to charge admission, creating the need for still more structural organization. Burning Man became an official corporation, with a board of directors, a budget and annual conferences. Still more rules were created. People were searched before they entered the grounds. Heretics who would not comply were banished from the camp. The annual city built up around the man grew to over 65,000 people. Missionaries were sent out to evangelize the world and plant congregations on every continent of the globe. In many ways, Burning Man LLC has become one of America's most successful mega-churches.

It is important to note that all this began with no religious intent on the part of its founders. If anything, the original Burning Man ethos strove to be anti-authoritarian on all fronts,

---

[42] Doherty, Brian.

including the religious. As we have seen, however, we can never control all the variables. Our move to self-sufficiency has made our freedoms too dangerous for us to handle, so we must try to regulate them by any means necessary. Lists of rules. Codifications of doctrine. Control of the money flow. Advertising. Evangelization. It's all right there in Nevada.

A skeptic might argue, however, that Burning Man actually did have religious undertones from the start. That first night on the beach, for example, was not any random night, but was chosen specifically because it was the Summer Solstice, a time of religious significance going back to the prehistoric era. As it progressed, the movement was very quick to appropriate the concept of the religious pilgrimage, and the spiritual trek into the desert is practically a religious cliché. Overtones of trying to recapture Eden were everywhere.

Even more significantly . . . just what was it that so captured the imaginations of the people that first night on the beach? Humans have been ceremonially gathering to watch people burn practically as long as there have been humans. As the Burning Man acolytes gather each year, they form a direct link back to the ancient Canaanites with their child sacrifices to Baal, the Aztecs and their sacrificial virgins, the heretics burned at the stake by the Spanish Inquisition, and the Salem Witch trials. Nothing is as cleansing to us as fire, and since the dawn of time we have gathered together as communities to ritualistically offer up burnt oblations to our gods. It is no wonder the first burn touched something so deep and primeval (dare we day *religious*?) in the watchers on the beach that night.

This is not to castigate Burning Man in particular. It didn't set out to become the Southern Baptist Convention. It's just that this is what we do. All of us. We start with the playful impulse, the spark of genuine insight, the leap of faith, the spontaneity of a beautiful moment—but all of them are too scary and unpredictable and must be tamed. Certainly we can't live our lives this *intuitively,* we tell ourselves. We prefer the safety of the systems we create, even when they end up, like Burning Man, promoting the very values we claim to despise. This is true of all of our religious systems, just as it is true of all of our technologies.

This is why, in the long run, it doesn't even matter that the Burning Man list of commandments "guarantees" freedom. It can't. The very idea folds in upon itself in incoherence. It's not the nature of the commandments that is the question, it is the fact that there are commandments at all. Just as with our tech devices, we seldom see the larger systems lurking behind what we do.

We can make some very safe predictions about how all this will go. Burning Man Asia (or Australia, or Europe, or Africa) will vie with Burning Man America for doctrinal and organizational supremacy. There will be a Great Schism, and many different denominations will form. This will be followed by a Reformation of sorts, as groups travel even further into the desert, trying to reclaim the spirit of the original Apostles. Burning Man T-shirts and kitsch will be sold online. Some of the new popes will be corrupt. The church will begin sponsoring candidates for political office. There will be sex scandals. Finally, the old Burning Man will be cast aside, and a whole new movement will start up (on a beach on Mars, perhaps?) trying to incorporate an even newer version of the same very old ideas.

MYTHS

The technological world also plays an important religious role in society by creating foundational myths about itself. In doing so, it co-opts not only religious language, but also a whole range of iconography and images designed to create worshipful attitudes. Examples include *The Myth of Modest Beginnings*, where a simple garage in Palo Alto or Sunnyvale replaces the stable in Bethlehem as the humble birthplace of the savior of the world, and the nearly endless references to *David and Goliath*, as small start-ups battle with their gigantic predecessors.

Consider the ways people have used elements of the Christ story to recite *The Myth of Steve Jobs*: Like Mary, Steve's mother became pregnant out of wedlock and had to travel to a foreign land to give birth. He was raised by adoptive parents, and quickly gained the favor of his teachers. When he began working, he gathered together a group of disciples (the Mac team) and in righteous indignation, set out to throw the money changers (IBM) out of the temple. He performed many miracles (the Apple II, the Mac) and was a teacher of great wisdom (the Stanford graduation speech). He was crucified (fired by Apple) only to be resurrected (re-hired by Apple). After his return, he performed more miracles (the iPod, the iPhone) before eventually ascending into the heavens. He left his church in charge of a trusted disciple (Tim Cook, as St. Peter) and the worshippers who mourned his departure pray to this very day that somehow his spirit will continue to guide the direction of the company.

The fact that this has very little to do with Steve Jobs, the actual human being, doesn't really matter. As with any story, it's not so much the details you put in as the ones you leave out that are so important. The mythical Steve Jobs is the god the people want to worship, not the domineering narcissist of reality. Why not give the people what they want?

THE SINGULARITY

It would be impossible to discuss the blurred lines between technology and religion without mentioning the Singularity. Few ideas have gained such pop cultural traction in the past twenty years, with references to Singulatarian concepts regularly appearing in movies, TV shows, songs and advertisements. This is despite the fact that huge leaps of faith are required to believe in it. Or it could be that those leaps actually account for its popularity, since belief in the Singularity allows a person to live within a religious framework, while at the same time eliminating pesky notions about things like God.

The Singularity emerges out of studies of the exponential rate of technological change. Let's go back to our example of the man living in the little village in the Middle Ages. This time, instead of making him a farmer, let's make him a brilliant inventor. Let's say that he created a simple machine that could fly. Since none of the other villagers could understand it, and since the village was isolated, when the inventor died, his "plane" died with him.

Today, the inventor could post the plans for his machine on the internet, and within seconds, people all around the world would have access to the design. They would then be able to improve upon it, or develop other uses for it that were not anticipated by the original inventor. These new ideas would then also be posted instantly, reaching a still greater audience and leading to still further and faster technological growth.

New ideas no longer have to be passed along slowly, like in the old days, via migration, trade caravan or as a result of warfare. Our ability to communicate instantaneously means that the rate of technological change now increases exponentially rather than in a linear fashion. With so many people working on the same ideas in so many different places, it is now possible for an idea to be technologically obsolete within minutes of its first conception. Certainly by the time any device is actually put into production, it is already far behind the times. Pity the poor people who camp out on the sidewalk all night to be first to buy the latest techie gizmo, only to walk out the next morning holding an object that is already obsolete.

In our scenario, the invention of the flying machine would have had virtually no impact on life in the little village. After the inventor died, things would have gone on peacefully as they had for years. For people today, the increasing rate of change is highly disorienting. People are fearful of what is going to happen next. Enter inventor and author Ray Kurzweil. According to Kurzweil, we don't need to be afraid of the future because we are about to reach a point, called the Singularity, when advances in artificial intelligence, robotics and bio-engineering will be such that the machines will finally be smarter and more capable than us. They will be able to learn, and will make decisions better and faster than we ever could. They in turn will develop technologies beyond our wildest dreams and travel to the ends of the universe, transforming it as they go. The Human Era will come to an end, ushering in the glorious Age of Machines. This "next stage in our evolution as a species" will start, says Kurzweil, by the year 2045.[43]

The exact way the details work themselves out depends on which Singulatarian you talk to:

- The trans-humanists picture a future world full of cyborg-like hybrids, where humans still exist, but their capabilities are "improved" by all kinds of technological enhancements. Our brains will be hooked up to super computers, for example, enabling us to walk around displaying all kinds of super computing powers. When our livers fail, we'll just go into the shop for a mechanical replacement, enabling us to live on forever like an old Pontiac GTO.

- Others picture a world where advances in our knowledge of cellular biology will allow us to live indefinitely by re-engineering our cells so that they never age. In this case, we will co-exist with the machines even as we cede control to them.

- Still others say that we will someday be able to upload our consciousness to computers, giving us a kind of softwarish immortality.

- Another possibility is that advanced artificial intelligence will be able to replicate the activity of the human brain so effectively that a computer will develop consciousness on its own. A robot with such a "brain" would then be a sentient being, no different from

[43] See *The Rise of the Spiritual Machines* and *The Singularity is Near.*

humans. A subset of this idea pictures a mass of super powerful interconnected computers so massive and super and powerful and interconnected that a whole new type of consciousness that we can't even imagine will evolve out of their collective activity.

• Finally, say the dystopians, it is possible that the machines will come to view humans merely as useless hindrances, and simply eliminate us altogether (these scenarios make for the best sci-fi movies).

Anyway you look at it, say Singulatarians, the Singularity will usher in the dawn of a new era of consciousness that will change the world forever. Sound familiar? *You will not die*, says the snake. *You will be like God.* The Singularity promises it all. Kurzweil takes over a hundred supplements a day, hoping, like Simeon, that he will live long enough to see the glorious day.

The Singularity's critics say this is all a bunch of nonsense. Even if technology continues to increase at an exponential rate, which is not a given, we still have absolutely no idea how to even describe what consciousness is, much less replicate it. Also, the more we learn about the human brain, the more complex it becomes. We are nowhere even remotely close to being able to mimic anything but its most rudimentary functions. And even if we could teach a machine to actually reason, there is no indication that reasoning itself is what makes a human a human in the first place. What about the emotions? And creativity? Just because a robot is able to mimic a person's behavior, it will not be possible to tell whether it has a genuine inner life. We have already seen that it cannot have a spiritual one. Since there is no indication that the essence of humanity can be reduced to mere algorithms, we may be ceding control to entities that are actually inferior to us. In that case, the Singularity will represent a move backwards.

When confronted with these skeptical arguments, many Singulatarians resort to a Technologist of the Gaps evasion.[44] "We may not understand these things now," they say, "but someday in the near future, some super duper smart technologist with one of his super duper smart machines will figure this all out. It's just a matter of time. We have faith. After all, at one point we didn't think it was possible to send a man to the moon."

This ignores the obvious fact that it actually *was* possible to send a man to the moon. If something is impossible to do, it won't be done, no matter how much time and how many super techies you throw at the problem. It's hard, however, for people whose minds have been filled with grandiose utopian visions to ever admit that they might be wrong, especially when that vision includes them becoming some sort of a god.

The joke may end up being on the Singulatarians. If the Utopian drive towards self-sufficiency is solely a human characteristic, the machines may see no reason to keep going and just shut themselves down. While this wouldn't make for a very exciting movie, it would be pretty funny. I guess we'll just have to wait until 2045 to find out.

---

[44] A frequent gambit of many people when they can't explain something.

"INDOCTRINATION"

PART TWO: PRACTICUM

"Improvement makes strait roads;
but the crooked roads without improvement
are roads of Genius."

William Blake

SOME THOUGHTS FOR PARENTS

Parenting is a tough job nowadays. In addition to doing what they have always done, parents now have the added pressure of trying to figure out how much technology their children should be exposed to. Just about every parent is worried about this, but no one is quite sure what to do. Parents, especially the parents of young children, are all looking to each other for direction, but in a world with so many mixed messages nothing seems very reassuring.

Here is the crux of the problem:

We know that the same things that are good for us are good for our children—communion with nature, other people, God, and living in freedom.

We are bothered by the impact our own technological addictions have had on our lives, and don't want this for our children.

We have always known that plunking kids down in front of screens is not good for them, but all the propaganda swirling around us is telling us to do that very thing. Many of our major societal institutions, such as our schools and churches, have completely sold out to the technological imperative. How are we supposed to trust our own parenting instincts, when so many of the "experts" seem to disagree with us?

The purpose of this section is to help clarify some of these issues. The bad news is that nothing is going to stop the technological onslaught. We are going to keep moving down the line of the continuum faster and faster all the time. The good news is that it still is possible for us to choose how to live our own lives, and how to raise our families.

Most of the warnings we hear about children and technology are half-hearted attempts at compromise. People are resigned to the fact that tech is here to stay, so the talk centers on how we are going to have to adapt our children's lives to it. Instead, we should start from the much stronger position of thinking about how technology may or may not fit our priorities.

Let's start, then, with a question that will help rid us of some of the clutter. It's a simple question, one that is probably even self-evident, but one that we seldom ask in the context of parenting:

> *As a parent or parents, what steps are you currently taking to prepare your children for the day that they are going to have to opt out technologically?*

This question completely re-orients the discussion, doesn't it? I would encourage you to read it again, and ponder it for a moment. For people of faith, parenting has always been a counter-cultural act. We have always understood the need to teach our children to carefully critique the culture around them and to understand where, when, how and why they should not conform to it. This has been especially true in times when the culture has promoted ideas, values and practices that are bad for children. Since we are living in one of those times, we are going to have to start making some tough choices. Here are some thoughts on how to navigate these waters.

One of the most important jobs of parents has always been to be the regulators of the flow of information. We are our children's primary gatekeepers to the world. From us, they learn how to view life, other people and themselves. Since children are still in the process of developing rational thinking and emotional maturity, we have always understood that the truths of the world should be revealed to them a little bit at a time, in psychologically digestible doses. Too much information too soon is simply impossible for a child to process, and can be very damaging. A five year-old, for example, doesn't need to know the same things about the world that a twelve year-old does, and a thirteen year-old's psyche might be shattered by the same knowledge that a nineteen year-old could handle with ease. This is particularly true when it comes to making sense of all the evils in the world. It has therefore been up to the parents, who know their children best, to carefully monitor the pace and content of the information flow.

In the past, parents got help with this from the larger community. The elders of the tribe, village or town developed important rites and rituals that publically marked a child's passage to adulthood. Step by sequential step, year by successive year, the children learned what they needed to know in order to become fully functioning adults. This was all done in a culturally cohesive manner, led by adults who, as preservers of the culture, were viewed with utmost respect.

All of this gets upended the second you hand a child a smartphone.

In an instant, control of the information flow shifts from the parents to the device, from the village to the world, from the elders of the tribe to God knows whom on the internet. There is simply no way a child is prepared to handle the psychological, intellectual and spiritual dangers she is about to face. The potential for harm cannot be overstated.

First, consider the content. All of the foulest thoughts from every darkest corner of the human mind are now available to our children on the internet, instantaneously, in the privacy of their bedrooms at night. Parents who would have never made this information available to their children in any other form, whether in books, films, magazines, videos and the like, suddenly now expect their young children to be able to wade their way through the polluted waters—by themselves and with no direction— without getting sick. It is simply not possible. Sadly, all it takes is one viewing of certain images or one exposure to certain thoughts for the scarring to be permanent.

Second, consider the connections. Not only are people posting all of this garbage online, but many of them would love to hand it directly to our children. We would never let these people into our homes, so why are we letting them into our children's bedrooms? It was only about five or six years ago that every child development expert was telling us that children's internet usage should be limited to the family computer, and only in the presence of their parents. Can the psychological needs of children have changed so much in such a short time? Not likely. Children simply cannot be expected to safely make their way alone through a world of manipulative adults.

Third, consider the conditions. No responsible parent would say, "What I *really* want for my children is to be immersed, alone and for hours each day, in an artificial environment—an environment that values immediate gratification, shallow thinking, and abbreviated communication, that conditions them to believe that commercial products are the key to life's happiness, that causes them to develop various addictions, and that slows their intellectual and spiritual growth." Yet this is precisely the environment that every parent who hands his or her child a smartphone provides.

The fundamental nature of children has not changed. They still need the world presented to them one small step at a time. Too much information too soon is still bad for them. Short of actual abuse, nothing has more potential to rob children of their innocence than the internet. They need protection from their devices, protection that must come from their parents.

Information overload inevitably leads to a loss of respect for the elders of the tribe, including the parents. It used to be that the oldest people in the village were the most respected. As keepers of the collective memory, they stored their village's entire history in their heads—everything that had come before, everything that had been learned, everything that needed to be known. It was crucial for the elders to pass this information along before they died, or it all would be forgotten. Thus, people worked very hard to make sure the transmission of information took place, and special care was taken to respect and attend to the elders as the carriers of tradition.

Now that the collective knowledge has been placed online, the elders have been robbed of one of their most important, time-honored tasks. Consequently, their role in the culture has been diminished. When a similar shift happened long ago with writing, the elders maintained their position by becoming the keepers of the books. The technological world today has no place for anyone or anything that is old.[45]

It's not like people today are going around saying we ought to disrespect our elders. They have been venerated for thousands of years, so some residue of that respect remains. It's just that the culture no longer needs them. Kids may still love their grandparents, for example, but if they want to know something about the past they're going to look it up on the internet, not sit down and listen to one of grandma's old stories.

This new way is far more efficient, but the personal connection has been lost. Sitting around a fire listening to an elder recite a tribal myth is not the same as typing searches into a box. Neither is taking the time to ask grandma about her childhood. Both involve active presence. And because the internet presents all information as having the same level of authority, each group's particular stories sit alongside all others, right next to advertisements for breakfast cereal.

All of this has implications for parents today. Childbirth classes never use these terms, but becoming a parent is one of the first steps toward becoming a tribal elder. As such, the parent becomes an authority figure who is worthy of respect.

---

[45] This is one of the reasons it's so hard for a person over fifty to get a job in Silicon Valley. The older people remember the days before any of the latest tech widgets existed, and understand that the world got along just fine without them. This does not mesh well with the messianic aspirations of the techie gurus, who require true belief and unwavering devotion from their employees. The older people have seen these messiahs come and go many, many times, and are just not as likely to bow down to them. Thus the very industry that needs the wisdom and perspective of older people more than any other ends up laying them all off.

"Honor your father and mother," says the Ten Commandments. "Children, obey your parents," says the New Testament. All traditional cultures have rules such as these. Parenting comes with a built-in authority, one that is necessary to do the job effectively.

The internet erodes all this. We try to teach our children to do one thing, then the internet tells them it's okay to do something else. We try to pass on our wisdom, and the internet tells them our ideas are stupid. We try to pass on our values, and the internet mocks the very ethics we espouse. What is a child supposed to think? No matter how much we know, we'll never "know" more than Google. The more exposure children have to the internet, the harder it is for them to view their parents as having any authority at all. All children eventually figure out that their parents are not omniscient, of course, but they need to be given a firm foundation for their beliefs and to be presented with a coherent view of the world. This has always been a difficult task. Giving children unsupervised access to the internet now makes it impossible.

Everyone wants a list. Preferably with bullet points. "What are we supposed to *do*?" they ask. I think it's important to ask a different question. Instead of asking what we should *do* with our children, we should ask who we want our children to *be*. The trouble with lists is that they too can become technologies. Just administer the correct seven steps and voila´— successful children! Life is seldom so simple. Asking who we want our children to be keeps the focus on the end game. Sure, we want our five year-olds to act a certain way now, but what we really want is for them to be certain kinds of people when they are adults.

When we were raising our kids, for example, we wanted them to be adults who were readers and deep thinkers who took their faith seriously. We wanted them to be kind and thoughtful, and to care about other people. We wanted them to work hard at whatever activities they chose to participate in, and to learn to make their own decisions with integrity. Above all else, we wanted them to know they were loved.

Keeping the end game in mind made our technological decisions much easier. Instead of asking, "How much television is okay for our kids to watch?" we asked, "How is television going to help us raise kids who are deep thinkers?" Instead of worrying about the content of certain video games, we asked how having them in the house at all would help them become people who were kind and thoughtful. Instead of just accepting the technological imperative, we decided which elements of it would be beneficial for our kids. Unsurprisingly, it turns out that few of them were.

Not that the process was perfect. Like any set of parents, we made our share of mistakes, many of them big ones. Through all the ups and downs, however, we think our kids are, with God's grace, well on their way towards being the people we hoped they would be. Let's cut through the clutter again: Parenting is difficult enough as it is. Would a steady stream of media input help or hurt you in your task? So many people fret about how hard it is to limit their children's access to technology. We feel that the choices we made for our family actually made the job of parenting much *easier* than it would have been otherwise.

No one has ever understood these principles better than the Amish. The Amish are not anti-technology. They just understand that there are systems behind every technology that have the power to threaten their values. Despite the fact that they drive around in buggies, for example, they do not think cars are evil. A car is, after all, just a device. What they figured out earlier than the rest of us was that the widespread adoption of *auto mobiles* would eventually lead to the end of their cultural cohesiveness, because people would be able to live further and further apart. Since communion with each other is one of their key values, they choose not to drive cars. The Amish person driving a buggy is not afraid of the future, or of machines, or of society. He or she is instead making a profound statement about the importance of family and community.

The tech world doesn't understand this at all, and is therefore fearful of their mindset. If the Amish are correct, it means that the entire technological enterprise is on the wrong track. This is why the media keeps up a steady stream of mockery about them, laughing at their clothes and hair and making sure to spotlight any members who are hypocritical or who have fallen away from the fold. The Amish couldn't care less. By taking the stance they have, they have a deeper communion with nature, each other and God, and have maintained the freedom to live their lives the way they please. We should all be so wise.

The time is coming when our families are going to have to opt out of the technological agenda for moral reasons. This will be easier to do if we have been preparing for it all along. Rather than unreflectively adopting each new technology as it comes along, parents should start teaching their children to say "no" when they are small and the choices are not so consequential. It is unrealistic to expect them to make wise choices when they are older if they did not learn how to make them all along the way.

Ideally, families of faith should find themselves falling further and further behind the technological curve the faster the pace increases.

So try keeping the Sabbath holy by keeping it tech free. Try a weekend-long family digital detox once a month. Turn off the devices at the family table, and lose the videos in the car, especially when you are on long trips and there is so much time to talk and think. Turn off your own devices, or don't buy them in the first place. The American Academy of Pediatrics recommends that children under two have virtually no access to screens. Psychologist Mari K. Swingle, who has extensively studied the effect of the internet on the teenage brain, recommends that contact with screens be severely restricted before the age of six. Some Waldorf schools raise that age limit to twelve. I think a child should wait at least until eighteen to have a smartphone.

If your children are already addicted to their devices, just saying "no" will not be effective unless you also teach them how to fill the empty spaces that are left behind when you take their devices away. As with any narcotic, they will have withdrawal symptoms. It is likely that their imaginations will have been flattened out by too much screen exposure, so they will have a hard time thinking up things to do on their own. So read to them. Talk with them. Listen to them. Have them play outside, or learn to cook or play an instrument. Most of all, spend time with them. The world is a pretty wonderful place and there are millions of things to do. What a pity it would be for our children to waste the only childhoods they will ever have with their noses stuck in glowing, addictive boxes.

A fun way to raise their awareness of technology's effects is by doing ordinary tasks using obsolete technologies. Have dinner by candlelight. Walk someplace together, instead of taking the car. Sit by the fireplace and tell stories. Even better yet, sit around a campfire. Our ancestors did these things on a daily basis and no doubt considered them thoroughly unremarkable. Today we like to go back technologically when we have something really special to commemorate—perhaps because it puts the emphasis back on what is human. Children should learn that there are different ways of doing things, and that their use of devices is not mandatory, but a choice.

This is not a list. These are just a few suggestions about the types of changes you can make. Each family should carefully and prayerfully consider these things for themselves, based on their own values and the things they want for their children.

One last thought. It will be a lot easier for you to do this if you have a group of like-minded people around you. The Amish, after all, have each other. Little Susie is going to be a lot happier if she knows that there are twenty-five other kids at church and school who don't have cell phones either. As a tribal elder, you too will be a lot more secure in your own choices if you have a group of fellow elders to lean on.

SOME THOUGHTS FOR YOUNG ADULTS

I hate the word "millennials". I especially hate it when older people use the term to label the entire age group as tech addicted, anti-social and lazy. First of all, as with any generalization, it is simply not true. The bigger problem for me is that those doing the criticizing are just as addicted to their devices as the young people they are disparaging. They are also the ones who caused the problem in the first place. After all, who made the devices and gave them to their children?

Unfortunately, even though the technological mess we are in is not their fault, it will soon be the millennials' problem. It will be an even bigger problem for the still younger generation (yet to be labeled and yet to be disparaged) that is in our schools today. The millennials, at the very least, can remember a time before everyone was saddled with a phone. The kids of today cannot. So what can a person do about it? Here are some thoughts.

One of the hardest things to do in life is to be authentic. As people like to say today, it's important to just "keep it real." I'm not sure we ever fully get there. Hopefully, as we get older, we get closer and closer to being able to think wisely, say the things we mean to say, and be the people we truly ought to be.

For most people, this is a long, constant process. We are not the same at twenty-five that we were at fifteen. We grow up under the influence of our parents and families, but as we get older, our experiences of the world, the input of the people around us, and the ideas we are exposed to all help us develop into unique individuals.

One thing that always made this process easier was that much of it took place in private. We would lie awake in our beds at night thinking about the world and our lives, or sit alone on a mountainside or a city rooftop, or go for long walks along the beach, taking the time to ponder, to be silent, to wander, to wonder.

In our private worlds, there was the freedom to make mistakes, to change our minds, to not always have to explain everything. We could try on one set of ideas one day, only to discard them the next for something completely different. We also had the privacy in which to develop our spiritual beliefs, since our meditative moments are often our most intimate. All this growth was accomplished without any external eyes constantly prying into our business. We could be vulnerable, at least with ourselves, without always having to worry about being judged for it. After wrestling with ourselves in private, we were then able to live as increasingly authentic people in the world around us.

All of this has changed radically in the past ten years because of technology. The results are disheartening.

Social media was supposed to be the ultimate vehicle for self-expression. After all, we could post photos of ourselves, tell the world what we liked and didn't like and express our opinions about anything and everything. What once promised to be so freeing, however, quickly devolved into a world where obscuring the truth became deliberate and strategic. A lack of privacy always encourages deception:

> *Make sure the background is interesting, check your shoulder angle, place your hand on your hip in a teacup pose, put on a duck face or a fish face, check your make-up, adjust the lighting so the shadows block out any imperfections, take the selfie, make any necessary adjustments, take fifteen more selfies with slight changes in each, choose the best one, crop out the bad parts, use a filter to get just the right vintage touch, then post online and wait for likes.*

Yep. *Just keepin' it real.*

Why all the pressure to be perfect? Because the world is watching, twenty-four hours a day. Every. Single. Day. Never before has a generation lived under such constant scrutiny. Every time we post something online, whether a picture, an opinion, or even a "like", we open up another little piece of our souls to be evaluated and commented upon by the public.

We, too, get locked in an endless cycle of making public judgments about others. The constant liking and un-liking, the posting and reposting of things we agree or disagree with, the need to state our opinions, even if they haven't been asked for—it's all so relentlessly narcissistic and judgmental. Even not responding to a post can now hurt someone's feelings, as can all the other new ways we have learned to publically shame others without necessarily saying anything outwardly cruel at all. Is it any wonder that we spend so much time developing deceptive personas?

Predictably, this inability to be authentic is beginning to take its toll on young people. The constant pressure to live up to some sort of unattainable ideal is causing many of them to crack. A recent study in the Harvard Review links the use of Facebook with negative overall physical and emotional well-being. The more you're on social media, in other words, the more unhappy you are.[46] Another recent study notes the alarming rise in depression among young people today, a trend that exactly correlates with the appearance of the smartphone.[47] Yet another shows how constant emphasis on physical perfection online is a key factor leading to a lack of sexual satisfaction among young adults.[48] *Have Smartphones Destroyed a Generation?* asks a recent article in *The Atlantic*.[49]

It makes perfect sense. Never before have so many people been so publically compelled to present themselves as always being beautiful and always being popular and always having fun. They can never let down their guard.

When these messages become internalized, the definition of what it means to live a good life becomes warped. A good life is now one where one person's artificially created image is able to superficially please the artificially created images of others. No wonder depression is becoming an epidemic.

And as to those places of refuge that the youth of every generation prior to this one have gone to for solitude and privacy, those places where we used to think things through, change our minds, and find grace for our mistakes? They are gone. Our devices are always with us, even in our beds at night. A cruel judgment always waits, only a click away.

---

[46] Shakra, Holly B. and Christakis, Nicholas A.
[47] Twenge, Jean M. *iGen*.
[48] Dovey, Dana.
[49] Twenge, Jean M. September, 2017.

So what can be done? The answer is simply this: think about who you are as a person of faith, and then allow into your life only those technologies that enhance your values. Do not sit passively by and let the culture make these decisions for you.

The first step is to become aware of the impact various technologies are having on your life. Try the digital detox and pay attention to the effect it has on you. Educate yourself. Become aware of the current research. Read. Make an honest appraisal of where you currently stand.

The second (and much harder) step will be to then apply what you have learned and take the countercultural steps necessary for you to maintain your authenticity.

Realize that your schools, churches, parents and friends have largely given themselves over to the technological imperative, and will not understand what you are doing. There will be resistance. I have no faith that my generation, except in individual cases, is going to be much help. We have already been down to the silicon crossroads and sold our souls to the technological devils. One generation makes the machines, leaving the ethical fallout to the next, all the while proclaiming their innocence of the damage done. It's a familiar pattern.

But maybe, just maybe, if you learn enough and read enough and talk enough, a growing number of people your age will grow tired of being manipulated, of having your brains experimented with, of being forced into addictions you did not choose, of having your spirituality, which ought to be as limitless as God, bound into little boxes—and say enough is enough. No person, and no generation, is obligated to accept the cards it has been dealt. The same is true for you. All you stand to gain is your freedom, and your authentic self.

SOME THOUGHTS FOR CHURCHES*

1. Since technology . . .

Is addictive,
Values efficiency above all else,
Marginalizes the elderly,
Abhors the past and disdains old things,
Hides its intentions,
Uses language deceitfully,
Grows unpredictably,
Keeps us from having communion with nature, each other and God,
Enslaves while promising freedom,
Constantly needs upgrading, and
Is an expression of humankind's self-sufficiency. . .

. . . perhaps it ought to be used sparingly and advisedly in a church setting.

---

*The next two chapters are geared toward a very specific audience, and may not be of interest to the general reader. Consider them as internal memos, if you will. Feel free to skip to *Arete of Cyrene*, if necessary.

2. After going through an entire week of looking at addictive screens, the last thing people need on Sunday is to come to church and see another one. They desperately need a refuge from the technological deluge. The church can play a pivotal role in the culture by providing one.

3. With implications for addictions, including technical ones:

> *"It is for freedom that Christ has set us free! Stand then, as free people, and do not allow yourselves to be slaves again."*
>
> *Galatians 5:1*

4. "No church today can get by today without having a website." Really? *Where is your god? Maybe he's sleeping, and you've got to wake him up*![50]

5. It is true that the new wine must be placed in new wineskins.[51] But the new wineskins are still wineskins. There are not some new technology. It doesn't say that the new wine must be placed in oaken barrels, or a Thermos.

6. The generations most susceptible to tech addiction are the young. Any church services geared towards them should therefore be as tech free as possible.

7. Our moment of greatest danger is when we start to believe that God needs our efforts. In the end God doesn't need

---

[50] See II Kings 18.
[51] See Matthew 9:14-17.

us at all: if we are silent, the very rocks will cry out.[52]

8. The twentieth century Church saw a massive influx of American marketing techniques. After two thousand years of apparent confusion, for example, it suddenly became possible to boil the entire gospel down to four "spiritual laws". It just so happened that they fit conveniently into a neat little booklet. And by putting the right bumper sticker on your car (*I Found It!*), or wearing the right T-shirt (*God Rules!*) your possessions could finally spread the good news for you. After so many years of this, is it any wonder that churches are having such a hard time resisting the technological imperative today?

9. They will know we are Christians by our love, not by our websites.

10. We would all agree that a church that helped foster a person's cocaine addiction would be morally bankrupt. So why is it okay for it to help facilitate a person's tech addictions?

11. Church growth techniques are always substitutes for listening to the Spirit. This is especially true of the ones that "guarantee" success.

12. Technology promises spiritual redemption through the use of material goods. This, by definition, is impossible.

---

[52] See Luke 19:40.

13. At the Last Judgment, people's names are written in the Book of Life.[53] Evidently heaven never got the memo that nobody reads books anymore. Shouldn't it at least be the *iPad* of Life by now? Why can't they keep up?

14. The tech you buy today is already obsolete. And just *how* big is your church budget?

15. "We should use all the technologies that are available to us in spreading the gospel."

    Jesus didn't.

16. It takes quite a leap to put one's faith in machines that no one can fully control or understand.

17. We're supposed to tear the false idols down.[54] Why are we now setting them up in our sanctuaries?

18. The church, temple, or synagogue that buys into the technological imperative will, without a doubt, gain economic, political, religious and social power, since these systems are all interconnected. It is, however, a Faustian bargain.

19. In his vision on Patmos, John sees a time when no one will be able to buy or sell without a mark on the right hand or forehead.[55] I have no idea what this

---

[53] See Revelation 20:15.
[54] See Deuteronomy 12:3.
[55] See Revelation 13:17.

means, symbolically, other than to say that this is now technologically possible.

20. We have become so accustomed to viewing all problems as having technological solutions that we impose technological solutions on what are actually spiritual problems.

21. Many churches have recovery groups for addicts. It's time to add tech addictions to the mix.

22. Technological adoptions should always come with an exit strategy.

23. "We all want progress, but if you're on the wrong road, progress means doing an about-turn and walking back to the right road; in that case, the man who turns back soonest is the most progressive."[56]

24. Ideally, churches should find themselves falling further and further behind the technological curve the faster the pace increases.

25. What is your church currently doing to prepare its people for the day when they are going to have to opt out technologically?

The opportunity is wide open for the church today that really wants to make a difference. People need a quiet rest for their souls, a place where there is genuine communion, where they

[56] Lewis, C.S. *Mere Christianity*.

can listen to God in stillness, without having to fight through all the technological dissonance. Imagine what would happen if all the young parents at a church joined together to raise their children device free. Or if the kids in the youth group decided to ban cellphones at their meetings. Or if a church had designated screen-free Sundays. Members of a church who had overcome tech addictions could have a forum for helping those who were still trapped. The elderly could once again be encouraged to give voice to their wisdom. The possibilities are endless. Rather than starting with tech as the underlying assumption and trying to adapt all ministries to it, the church should keep God as the basic reality, using machines only when necessary and only if their harm is limited. If we can get this right, the church would be an oasis.

All by doing *less*.

SOME QUESTIONABLE ARGUMENTS

I don't quite understand an argument that is widely making the rounds today. One version of it goes something like this:

1. God told Noah to build the ark.
2. In order to build the ark, Noah needed to use tools.
3. Therefore, God wants us to use technology too.

Other versions of this argument utilize the story of God telling the Hebrews to build the temple, and the fact that the cross was made with a hammer and nails. My problem with this is that I'm not sure how God instructing one specific person or group to do one specific job for one specific purpose in one specific instance means much more than just that. Now, if God comes to you today and specifically asks you to build an ark, you'd better build an ark. But this would have nothing to do with the rest of us. Or even you, necessarily, after the ark was finished.

Genesis says that Noah was chosen because he was *a righteous man, blameless among the people of his time.* This proves the story has nothing to do with me! I also have a hard time seeing a correlation between Noah working outside on a plain somewhere all day, talking directly with God and doing the one unambiguous task set before him, and me then using the story to justify donning a set of VR goggles.

Besides, I think this sort of thinking could set a dangerous precedent in biblical interpretation:

1. God told Hosea to marry Gomer.
2. Gomer was a prostitute.
3. Therefore, God wants us to be with prostitutes too.

Doubtful.

Here's another version of the argument I have come across:

1. Jesus was a carpenter.
2. Carpenters use tools.
3. Therefore, to be Christ-like means to use technology.

One could just as easily say:

1. Jesus was a carpenter.
2. Jesus quit being a carpenter when he started his ministry.
3. Therefore, being like Jesus means leaving technology behind.

Perhaps what we should leave behind is our tendency to turn *every* story into a proof.

Another argument that seems questionable to me is what is called "The Cultural Mandate". This idea stems from the King James translation of Genesis 1:28:

> *And God blessed them, and said unto them, be fruitful and multiply, and replenish the earth, and subdue it: and have dominion over the fowl of the air, and over every living thing that moveth upon the earth.*

The argument goes something like this:

> *"The first phrase, 'be fruitful and multiply,' means to develop the social world: build families, churches, schools, cities, governments, laws. The second phrase, "subdue the earth" means to harness the natural world: plant crops, build bridges, design computers, and compose music. The passage is called the Cultural Mandate because it tells us that our original purpose was to create cultures, build civilizations--nothing less."* [57]

This one verse, and particularly the words translated as *subdue* and *dominion,* has also been used over the years to justify a wide range of other beliefs and behaviors, including but not limited to: the destruction of wildlife and their habitats, the idea of Manifest Destiny, the advantages of the Protestant work ethic, the hoarding of wealth, the establishment of the death penalty, and even as an argument to explain why aliens cannot possibly exist![58] This is an awfully heavy load for any one verse to carry.

---

[57] Pearcey, Nancy.

[58] Because humans are the ones who are supposed to be fruitful, of course, not aliens!

The trouble is, it cannot be removed from its context. As we have seen, in Eden there is nothing for the man and the woman to subdue and nothing for them to have dominion over. "Creating cultures and building civilizations" are the types of things they will later be *forced* to do as a consequence of their actions. As Paul goes on to say, creation was condemned to lose whatever purpose it had.[59]

Sadly, lurking under the good intentions of every attempt at Christianizing society is the same old Utopian impulse. This is why Jesus said to *give to Caesar the things that are Caesar's, and give to God the things that are God's,*[60] and continually stressed that his kingdom was not of this world. The two are mutually incompatible. There is a difference between a group of senators meeting together informally to pray for the country, for example, and the establishment of a National Prayer Breakfast.

Besides, which version of Christianity should take precedence? Each of us has a fragmentary understanding of God. The societal imposition of my theological viewpoint would inevitably rob you of the freedom to pursue yours. This goes for technology too. I do not have the right to drag you down the line of the continuum. Even if the Cultural Mandate were a real thing, no one would be entitled to enact it.

So we continue to create cultures, use technologies and build civilizations, because, well, that is what we do. From time to time God may interrupt our trip down the line to commission an ark or a cross or even a governmental authority,[61] but for the most part, he leaves us to our own

---

[59] Romans 8:20

[60] Mark 12:17

[61] e.g. Pontius Pilate, Nebuchadnezzar, Nero (Romans 13)

self-sufficient devices. We ought to be very careful about claiming that our actions have his mandate.

As to the existence of aliens, I respectfully refuse to speculate.

ARETE OF CYRENE

HOST: Welcome back, everybody. The show is just about over, but it looks like we have time for one last caller. Hello, caller!

ARETE: Hi! Thanks for taking my call. This is Arete of Cyrene.

HOST: You mean Arete the Cynic philosopher who wrote forty books, was the teacher to over one hundred other philosophers, and was reputed to have the oratory skills of Socrates and the beauty of Helen? What a way to end the show!

ARETE: Yes, it's me.

HOST: I'm just glad to be talking with a Cynic, not another one of those skeptics.

ARETE: Thanks! The old boys do tend to get a bit tedious at times. All they do is sit around arguing. And watching movies. Which they then argue about.

HOST: I'm glad you found time to call. What can I do for you?

ARETE: As a Cynic I agree with a lot of what you're saying, especially as it relates to living simply. But there is one question that has been nagging at me the whole time: Aren't you getting a bit too worked up about all this? I mean, change is constant. People have always worried about how technology was changing things, yet we've always survived. Plato was worried when writing was introduced. People had their concerns about the telegraph, the telephone, and television when they first came around. How is this era any different?

HOST: Ahh. The Elvis Argument. It gets trotted out every time. Do you know who Elvis was?

ARETE: Yeah, I heard his name once when the guys were watching some cheesy movie. I think it was *Blue Hawaii*. What does he have to do with anything?

HOST: People like to talk about how the society of the 1950s was irrationally afraid that Elvis's hip shaking was corrupting the youth of America. Then they say something like "See? Here we are today, doing just fine. Wasn't all of their concern oh so quaint?"

ARETE: What does that have to do with technology?

HOST: Nothing, really, but you hear it used *all the time* in arguments about tech. I guess people are trying to make the point that while we have always worried about the future, we've survived anyway.

ARETE: I don't see what a cultural adaption to Elvis's hips has to do with me confronting technological systems. If I once survived the measles but now have cancer, so what? What good does it do me to say I survived the measles? Still, the fact does remain that we *have* survived many technological changes.

HOST: But is surviving the same thing as thriving?

ARETE: Many people would argue that technology *has* made us thrive.

HOST: It has in many ways, but once again it's only half of the story. If our focus stays on the devices or methods in front of us it can look like we are thriving, but once we see the systems behind them, we become aware of the damage that has been done, especially to our spirituality.

ARETE: Kind of like gaining the whole world but losing your soul?

HOST: Exactly. It reminds me of a passage in *No Country for Old Men*. Have you read it?

ARETE: No, but the boys watched the movie. They were worried that the title was potentially micro-aggressive.

HOST: Really? *Very* interesting. In one particularly chilling scene Chigurh the assassin asks Wells, "If the rule you followed led you to this, of what use was the rule?" [62] He then shoots him in cold blood. In the context of our discussion the question becomes, "If the path we have taken technologically has led us to the point where it is killing us spiritually, of what use was the path?"

ARETE: Okay. But a lot of good has happened along the way.

HOST: At what cost? When Elvis first shook his hips, the girls in the audience were so over-stimulated they practically lost their minds. The trouble with constant overstimulation, however, is that it eventually leads to numbness. It's like any other drug. We survived Elvis's hips, but only because our sensibilities became so deadened that it took greater and greater stimuli to get any response out of us at all. The girls who went crazy for Elvis had a far greater capacity to *feel* than we do today, and were far more sensitively attuned to their environments. So were Elvis's critics at the time. Today, our entire experience of the world—of nature, each other and God—has flattened out due to technological over-stimulation. In our numbed-down state, our only recourse is to mock the people of the 1950s, just as we mock the Amish, and the ancients. Perhaps we are the ones who should be pitied.

The Elvis/Tech Argument depends on a couple of problematic underlying assumptions:

1. That the pace of technological change can keep on increasing without doing us any harm.
2. That all types of changes are equivalent.

---

[62] McCarthy, Cormac.

ARETE: These assumptions aren't true in any other field, so why do people use them here?

HOST: They shouldn't. Let's talk about what happens when we increase the pace, for example. Every kid knows that if you spin around on the playground twirly whirl fast enough you're going to fly off.

ARETE: I don't think every kid knows that. They've gotten rid of the twirly whirls. And the teeter-totters. And the really big slides. Even playtime at recess has flattened out.

HOST: It's sad, isn't it? I hope the audience still knows what we're talking about. Technology is like the twirly whirl. It has sped us up to the point that it is flinging off all sorts of things: reading, reflective thinking, talking with each other, communion, freedom, spirituality. It's a long list. And it's a high price to pay.

ARETE: Not to mention the fact that when the twirly whirl goes around too fast, you get dizzy and sick.

HOST: Right. We're seeing the signs of that everywhere in the damage that is being done to people's physical and mental health. What remains to be seen is how much faster it can spin around before it jumps off of its base completely. We cannot keep speeding up indefinitely.

ARETE: What about the second assumption, that all types of changes are the same?

HOST: It's problematic too. There's a famous anecdote about the techie guru who responded to someone questioning one of his technologies by retorting, "Well, you're wearing glasses aren't you?"

ARETE: Was that supposed to be profound?

HOST: No comment. I think it was a rather heavy-handed way of trying to insinuate that all technologies are the same, and that the glasses wearer would be hypocritical to not adopt the guru's latest product.

ARETE: But aren't you the one who has been saying that *all* technologies move us down the line of the continuum?

HOST: They do, but not in the same ways or nearly to the same degree. Eyeglasses help people see more clearly by restoring a function originally intended by nature. That is a far cry from being addicted to a glowing box created by some techie guru with a profit gleam in his eye. The assertion that all technologies are the same is just another rhetorical trick technology uses to hide its intentions.

ARETE: So what is different about the changes that are happening now?

HOST: In one of my science classes in school, we did these experiments with bacteria. The first thing that went into the petri dish was the agar, the growth medium. Then the bacteria were introduced. If the agar wasn't just right, the results of the experiment might be skewed, or the bacteria might not grow at all. Today, we are taking people's brains and dropping them into the corrupted agar of the screen-centered world. The result, especially for our kids, is that our brains are growing differently, and we have no idea what the long-term damage is going to be. So far, the short-term results have not been encouraging. If I were a kid today, I'd be highly resentful of my mind being experimented with in this manner.

ARETE: Are there any other ways the type of change is different now?

HOST: Sure. They keep telling us that since we were once able to shift from an agrarian economy to a manufacturing one, we'll have little trouble moving from manufacturing to the new tech economy.

ARETE: What's wrong with that?

HOST: It's not an equivalent shift. When people left their farms for the factories, the skill requirements actually went down. Anyone who could plow a field could put the right front wheel on a Model-T. In the move from manufacturing to tech, people's basic personality types have to change, since the tech world only comfortably accommodates a particular kind of person. We can't force everyone else to do this kind of work.

And contrary to popular belief, the schools are not going to be able to train people for this. Imagine the poor kid at school who has a gift for poetry being forced to sit and write software code for hours on end.

ARETE: As if the sole purpose of words is to get machines to function! Who would do this to a child? It's demeaning. And demoralizing. I would have left school rather than suffer through it.

HOST: Me too. It's one of the many, many reasons why the techie gurus cannot be allowed to set the agenda for our schools.

ARETE: What about all the other jobs that surround tech, like customer service, sales, marketing, and the like?

HOST: These are the very jobs that are already starting to be eliminated as the machines become "smarter".

ARETE: So where does all this leave us?

HOST: Trapped.

ARETE: Thanks a lot.

HOST: There really is nothing we can do. As we have said from the beginning, we have to use technology in order to survive in the world, but every time we do, we move ourselves further away from God, and our freedom.

ARETE: Are you saying there is no hope?

HOST: No, there actually is a lot of hope. The first thing to understand, however, is that our basic problem is not technological. It's not political, religious, economic, or governmental either. It's spiritual.

ARETE: If the problem is spiritual, what should we do about it?

HOST: Nothing.

ARETE: So, we're right back here again.

HOST: No. Nothing is what we should do. There is nothing we can do to save ourselves, so we should do nothing.

ARETE: What about trying to be a good person?

HOST: Being a good person is very important, of course. Acting morally makes things a lot better for those around us, and it means our own lives will be filled with a lot less chaos, but it will not solve our basic problem.

ARETE: So, do nothing? I'm still needing a little help here.

HOST: The issue is not what we should do, but who we should be. And who are we? We are the children of God.

ARETE: What does it mean to be a child of God?

HOST: It means that we have inherent worth. That our lives matter.

ARETE: But how does a person live this out?

HOST: The best example comes from Jesus himself. The snake's offer to the woman is *You can be like God.* The book of Philippians says that Jesus, by contrast, did *not consider* *equality with God something to be grasped*, but instead humbled himself and did God's will.[63] That's it. Being a spiritual person means not grasping for equality with God. It's the opposite of self-sufficiency. Micah 6:8 says that we should *do justice, love mercy, and walk humbly in fellowship with God.* Jesus says we are to *love God and to love others as he loved us.*[64] And when we mess up, as we inevitably will, God still offers us restoration and forgiveness as his free gift—if we will only ask.

ARETE: It doesn't sound very complicated.

HOST: It's not. It is simple enough for a child to understand. We've made things way too complicated by turning these simple ideas into big systems.

ARETE: Are you talking about religion?

HOST: Religion can be just as much of a system as anything else. Too often it promises us communion with God, but doesn't tell us that we'll also be caught up in a thicket of theologies and egos and rules and structures from which there is no escape. All of it—religion, politics, government, economics, technology—it all stems from the same self-sufficient mind-set.

ARETE: But doesn't the move toward self-sufficiency *mean* freedom? If we are making our own choices, aren't we free?

HOST: If we are truly at our best when we are in communion with God, and if our self-sufficient moves always end up

---

[63] Philippians 2:6
[64] Matthew 22:37-40

enslaving us, that's not really freedom, is it?

ARETE: It is if you co-opt the word "freedom".

HOST: Believe me, it has been done. "Freedom" might be the single most co-opted word of all time. It's not a matter of making more lists of rules, creating new systems or even of reforming the old ones. It's a matter of recognizing who we are and stepping out in faith each new day with that orientation in mind. It is only then that we will experience all the things that are truly good for us.

ARETE: In a lot of ways that actually sounds a lot harder to do than following a list of moral or religious tenets or getting swept up in some political movement.

HOST: It's not necessarily hard to do. It just requires a lot of courage. There's a reason why the leap of faith is called a "leap". What makes it hard are the daily battles that we have with our own human nature and with the systems around us that are trying to distort our vision and rob us of our freedoms.

ARETE: You talk a lot about keeping an eye on the end game. How is this all going to end?

HOST: Let's go back to the continuum. People of faith believe that in the end there will be a restoration with God. This changes our diagram. Let me show you.

ARETE: I can't see the diagram. We're on the radio.

HOST: Oops! You're right. Sorry. I guess I'll have to describe it to you.

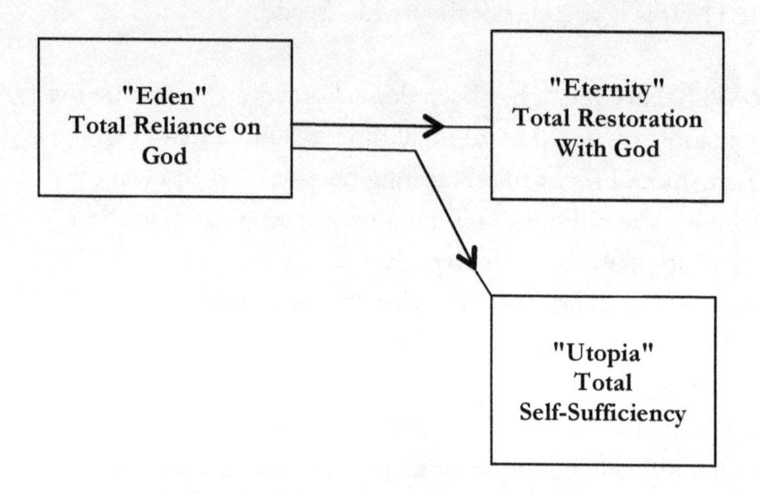

For most of human history, the two lines were very close to each other. People used simple technologies, but it was very apparent to them that they were still really dependent on God. This made it much easier for them to take the necessary leaps of faith. Today, as we cut ourselves off from listening to nature, each other and God, the two lines are rapidly diverging. The growing separation now enables us to believe almost completely in the illusion of self-sufficiency.

ARETE: Does that mean we shouldn't use technology at all?

HOST: It's not possible. We can't go back to Eden. There is no way for us to escape from the various systems we have been born into, technological and otherwise. At the same time, we should be very wary of each step forward into self-sufficiency. We cannot serve two masters. We cannot love them both.

ARETE: There are going to be some people who agree with your thesis—and want to spur technology on all the more as a result.

HOST: How so?

ARETE: They'd like nothing better than to get rid of the spiritual altogether. All of this "god" mumbo jumbo does us a disservice, they say, and keeps us from getting on to the tasks we ought to be doing to save humanity. The bigger the gap between the two the better, as far as they are concerned.

HOST: They do have a valid point. A lot of evil has been done by people claiming to speak for God. But abandoning the Spirit has never worked. The pathway to Utopia has always been littered with the bodies of its victims. But other people are not the enemy. Our true enemy is the hubris that *The Garden of Eden* says lies within each of us.

ARETE: How are you doing at living all this out personally?

HOST: Okay, I guess. Better all the time. I get to hide behind the anonymity of my radio show, so people don't really know who I am. The person I'm really worried about is the author of this book. I know for a fact that he doesn't find being a spiritual person particularly easy, and that he's mostly writing to remind himself of the things that are really important. He's also fully aware that this book is a technology too, and that he might just be spitting in the wind.

ARETE: Is there any hope for him?

HOST: There is always hope. Sometimes it's the only thing we have left.

Well, that's the end of our show today, folks. Thanks to all of you listeners, and a special shout out to all of you skeptics and cynics out there—you always have a special place in our hearts. Where would we be without you?

Arete of Cyrene

PART THREE: EXPRESSION

"Tell all the truth but tell it slant—
Success in circuit lies
Too bright for our infirm Delight
The Truth's superb surprise."

Emily Dickinson

HOW TO READ LIKE AN "F" STUDENT

Hi! My name is Madison and I'm, like, a freshman in high school. When I found out my school was going to be giving each of us an electronic tablet this year, I got so excited! Finally—something fun to do at school! My parents have always been way too strict with me, and are always monitoring the stuff I watch online and on TV. I swear, sometimes it's like living in a prison.

When I told my dad about what the school was doing, he wasn't too happy about it. He told me about a study he had read recently. It turns out that when people read online, he said, their eyes follow an "F"-shaped pattern.[65] They read the first couple of lines of a webpage all the way across, forming the top of the "F". Then their eyes shift over to the left, travelling downward. Somewhere around the middle of the page, I guess they get worried or something, because they slow down to make a second horizontal movement halfway across. This forms the middle bar of the "F". From there, their eyes go back over to the left and they quickly skim all the way down to the bottom.

---

65 Nielsen, Jakob.

"Can't you see what a problem this is going to be?" asked my dad. "Since you'll be reading online, you're going miss most of what's on each page."

"Don't worry, Dad. I'm pretty tech savvy. I can handle it," I said. In my mind I was thinking, "Hallelujah! This is going to save me, like, a *ton* of time!"

It turns out I was right. In my English class, we started the year out by reading poems. The first one was really stupid. It was about some dumb horse stopping in the middle of the woods. I mean, *who cares*? My teacher said that no one really knows exactly why the horse was stopping, or what the poem was really about. Why should I be forced to read the whole thing, then? Luckily, I didn't have to. Since it was on my tablet, I just used the "F" method. Here's what it looked like to me:

> Whose woods these are I think I know
> His house is in the village though;
> He will not
> To watch
> My little
> To stop
> Between
> The darkest evening of
> He gives his harness
> To ask if there is some
> The only
> Of easy
> The woods
> But I have
> And miles
> And miles

When I told my Dad about it at dinner that night, I thought he might praise me for being so resourceful. Instead, he just groaned and said that some old dead guy named Robert something or other was "no doubt rolling over in his grave."

"And just how does that affect *my* life?" I asked, as I got up to do the dishes.[66]

We also have to take a Bible class (I go to a private Christian school). It's usually the worst class of the day. Our teacher is an old prune who starts to get nervous if her blouse isn't buttoned all the way to the top. She told us that on the first night of her honeymoon, she and her *betrothed* (yes, she actually used that word!) spent the whole night in prayer, devoting their relationship to God. When Josh asked her what they did the *second* night, she gave him detention for a week.

Anyway, today we were supposed to read the first chapter of John. At first I went to one of those Bible websites where you can read all the different translations. After my great experience with the poem, I was ready to use the "F" method again. The trouble was I couldn't see the verses at all. The top of the page consisted of links to social media sites and a big banner ad for a company selling vacation packages to the Holy Land. When I started down the left side, a big pop-up ad slid out, offering me ten thousand free Bible study tools. I just laughed. I didn't want *any* Bible study tools, much less ten thousand! This wasn't getting me anywhere.

Then I had a brilliant idea. Why not just use the "F" method on a regular page from a real Bible? So I took a Bible off the shelf in the room and read:

---

66 The poem being discussed here is Robert Frost's *Stopping by Woods on a Snowy Evening.*

In the beginning was the Word, and the Word was with God, and the Word was God. He was with God in the beginning. Through him all things were made; without him nothing was made that has been made. In him was life, and that life was the light of all mankind. The light shines in the darkness
There was a man sent
witness to testify
was not the light;
The true light that
and though the world was made through him,
that which was his own, but his own did not
those who believed in his name, he gave the
born not of natural descent, nor of human
The Word became
of the one and only Son,
John testified concerning
said, 'He who comes after
fullness we have all
through Moses; grace and
is in closest relationship.

Our teacher said this was a pretty important passage and that we should read it carefully. By using the "F" method I think I was able to get the gist of it: that words are important and we should believe in God and Moses and all that kind of stuff. Besides, at youth group on Wednesday night they're showing us a Jesus movie. I'm sure it'll fill in anything I might have missed.

So, while my dad is not happy about all this and keeps ranting and raving about the state of education in America today, I have to say I love it! I have *so* much more time for shopping online and hanging out on social media. At the rate things are going, pretty soon I'm going to be, like, the best "F" student in the school! Isn't that great? Technology is the *best!* Praise God!

A BIOGRAPHY

He always wondered why the other kids were so cruel. When he would come to, he would find that they had been yelling at him. His mind had been far away, roaming among the stars, but when he came back to earth someone was always clapping their hands in his face. "Hey! Wake up!"

He'd always known that something wasn't quite right, that he didn't fit in. Couldn't they see? He was just trying to help them. There were so many things that were wrong in the world, so many things they could be doing better. But they didn't want to hear it, especially from him.

It wasn't any better at home. He couldn't believe that the aliens living in his house were actually his parents. His father was an accountant, and his mother worked at a flower shop. They loved him, but they didn't understand him either. Sometimes he felt so alone that he wanted to make the world disappear. So he would take off to outer space in his mind, a place where he had no limitations. There, he was finally free. When he'd wake up, there'd be his mom, or his sister, or another kid, yelling.

School was a torture chamber. He was easily the weakest member of the herd. In elementary school he spent recess by himself, walking alone by the fence at the far edge of the field, only to find that stuff in his desk was missing when he got back. At noon, he'd go to his cubbie and see that his lunch had been stolen. He spent each day in isolation.

When he got home, his mom would ask how his day had gone, but she already knew the answer. If she pressed him too hard, he would fly into a rage, still too raw from the daily round of insults to want to talk about it.

So he'd retreat to his room and play his role-playing games. There, he was a king, a warrior, a warlock. His life mattered. He was somebody. Best of all, in the games he could control all the variables. If someone was tormenting him, he could be ignored, or even eliminated.

The world outside was unpredictable. He didn't understand it. But the games were perfectly rational. If you made one decision, you would find yourself exploring a magical cave. If you made a different one, you could conquer powerful cities. You were rewarded for your valor, your intelligence, your ingenuity.

He played every day. As he got in deeper, level by level, he determined that this was where he wanted to stay for life. He found it harder and harder to tear himself away for dinner. His mom, not wanting to upset him, finally started leaving his plate outside his closed door.

In high school things got worse. One kid stuffed him into a locker. Others would trip him when he was walking down the hall for no apparent reason other than their own amusement. His homework was taken out of his backpack and passed around the class. The other kids would copy off his work, and then return the papers to him crumpled, ruined.

When he came home one day with a black eye, his mom asked him what had happened. "Nothing," he said, and went off to his room to fight dragons.

He did like some of his classes, especially Math. Math was safe. The numbers were always perfectly logical. They never let you down. If a problem was difficult, you just kept at it. If he thought it through long enough, he would always find the solution. Other kids would chew the erasers off their pencils in frustration, but he'd just go to that special place in his head and wait. The numbers would float before his mind's eye and re-arrange themselves into the answer as he watched.

He awoke one day in class to find that the teacher had been talking to him. "Two hundred and forty-seven!" he blurted out, loudly. It was the answer to the problem in his head, but not to the question she had asked. The other kids just laughed, and began hatching another plot for after school abuse behind the bleachers.

His math teacher, though, recognized his gifts. He felt like she was the first person who had ever really understood him. She started giving him special puzzles and games and separate problems to solve. Sometimes he'd go in and talk to her after school. While he did genuinely enjoy this, he was really just postponing the ambush that was waiting for him.

One day she handed him a book. "I thought you might be interested in this," she said. It was a book on writing computer code. He took it home and promptly devoured every page. He started writing simple programs out on paper. He didn't know whether or not they'd work, but he found the whole concept fascinating. He began begging his dad for a computer.

This was something new. The games were fine, but they were someone else's universe, the product of someone else's mind. If he had his own computer, he'd have a chance to create the perfect world he wanted, a world made in his own image. He came home on his birthday to find a computer on his desk. It had a large red bow on it, with a card from his parents.

That was it. He was gone. It was his world now. The little programs he had been writing actually worked when he typed them into the machine. He wrote a simple one to help his mom balance her checkbook. He developed a spreadsheet to help his dad keep track of his tools in the garage.

As he shared these things with his parents, the ice between them began to thaw. They were delighted with his creativity, and they told him so. For the first time, he realized that if he could make things that people liked, they would like him in return. With his computer, he could build a bridge between himself and the outside world. He could also retreat when he wanted, and no one would care. His mental flights started going deeper and deeper into space.

One morning during his senior year, he happened to find a book on Eastern mysticism in the school library. He scoffed at it. He had left religion behind a long time ago. What kind of God could possibly create this horrible world that tormented him every day? Besides, now that he had his computer, he no longer needed to be so submissive—God was definitely out.

When he started reading the book, however, he was astonished. As the author described the dreamlike states that were possible during meditation, it all seemed very familiar to him. "This sounds like what I do all the time," he thought. The book went on to describe not only flights outward into space, but even more intriguingly, trips into the innermost self, where god was. "Maybe there isn't anything wrong with me at all," he thought. "Maybe I'm just spiritually advanced. Maybe I'm some sort of god already."

When he got home that day, instead of working on his computer, he tried meditating. He followed the instructions in the book exactly, but nothing happened. Ordinarily, he'd go into his "zone" at unpredictable times when he was thinking about some other idea. Now that he was consciously trying to manufacture his thoughts, he couldn't do it. He kept getting distracted. He didn't have the self-discipline. This went on for several weeks. Finally, he realized that in order to get it right, he was going to need some help. He walked into the kitchen and asked his parents if he could take a gap year instead of going to college right away. "Where do you want to go?" they asked. "India," he said.

He got back from India utterly convinced of his own divinity. It turns out he had underestimated his abilities. He had quickly mastered all the lessons his guru taught him. In fact, it wasn't long before he had been able to accomplish meditative feats far greater than many of his guru's long-time disciples. After all, he had been already practicing for *years*. He quickly grew bored and restless. "If this is what being god is," he said to his master, "then I'm already there." He left two months early, secure in the belief that he was destined to do something great. He went to college that next year on a self-reverential high.

His roommates didn't know what to make of him. When he wasn't in class, he spent most of his time meditating. When he came to, he would make bizarre pronouncements that would leave them shaking their heads. "Astral projection through nuclear fission!" he would exclaim, and run out of the room to get lunch. He didn't have many dates.

Prior to going to India, he had always thought there was something wrong with him. Now, he wanted to find out why he was so far ahead of everyone else. He decided to major in Psychology. In his classes, he learned how people could be conditioned into predetermined reactions, thus bypassing their free will. He learned how variable reinforcement ratios could be used to foster dependency. He learned that people had an innate tendency to project human traits on to inanimate objects. He learned that because people were relentlessly egocentric, they were endlessly exploitable. He filed all these ideas away in his mind.

One day in his doom room, he awoke from his meditation session and screamed, "Massive Dharmic communicative networks!" and jumped up. His roommates just stared at him. They had seen this act before. This time, however, instead of just going to lunch, he sat down at his computer and began to code.

He coded for six days straight. He didn't sleep, didn't eat, didn't take a break. His eyes got red and bleary and his sunken cheeks looked like a skeleton's, but he kept typing, rocking back and forth at his desk, eyes riveted to the screen. His roommates offered to get him something to eat, but he just muttered to himself and kept going. He was on a mission.

By the end of the sixth day, he was finished. He sat back in his chair, satisfied with his work. "It's really good," he whispered to himself. He uploaded his new program to the internet, and fell asleep on his bed for twenty-four hours. When he woke up it was nighttime. One of his roommates was snoring. He walked over to his computer and looked at the screen. He was astonished to see that his program already had forty users. By the time he walked to the 7-11 on the corner and got a sandwich, the number was up to sixty-six. When he got back to the dorm, it was seventy-seven.

Soon it was up over a hundred, then two-fifty, then five hundred, a thousand, a million. He dropped out of college and moved to Silicon Valley. The empire he had been fantasizing about since childhood blossomed up around him.

At work, he was a tyrant. Now that he had the power, he didn't hold back. The poor programmers who worked for him didn't realize that his tirades were not personal. His standards of excellence were just higher than theirs, he told himself. Besides, he was entitled to the role reversal; he had been hurt all of his life. They didn't know that they were paying for the sins of his childhood tormentors.

He hired pretty girls just to sleep with them, and then fired them the next day. He always sent a pink rose to their desks the next morning so their humiliation would be complete. He began dressing better, and paying attention to his personal hygiene. "What kind of car would God drive?" he asked himself. He went out and bought a Ferrari.

Outside, in the real world, people began running their entire lives through his program. Everything they bought, everything they sold, all of their business transactions, all of their personal relationships, their sex lives—it all went through him. He designed a small box so that people could carry his program around with them everywhere they went.

It was like crack cocaine. People checked their boxes several hundred times per day. They neglected their spouses, their children, their work. They slept with the little boxes under their pillows at night.

Time magazine came to interview him for a cover story. "You facilitate the communications, the finances, the personal lives of virtually everyone on the planet," the reporter said. "What does that make you?" He was politically savvy enough not to take the bait. "Just a CEO," he lied.

People began asking for his opinion on all sorts of subjects—even ones he knew nothing about. Education, politics, medicine, religion, it didn't matter. He could pontificate on anything, knowing someone would be right there, fawning over every word. Others began to deify him, and took pilgrimages to his childhood home and to all the places he had visited on his travels to India.

And yet, it wasn't enough for him. There were still more people out there to reach. There was also the next generation to think about. He began to worry that his legacy would not outlast him more than a few years. After reading about the great nineteenth century industrialists, he decided to use philanthropy to immortalize his name and consolidate his power. The more lavish his donations, the more his fame grew, and the more outlandish his promises became. He would end war, disease and ignorance, he claimed. The words were preposterous even as he said them, but he knew that no one would question him. He had become a master at distorting reality and bending people to his will.

One morning, he stopped to get coffee at a little local shop in downtown Palo Alto. When he walked through the door, he got a total shock. There, sitting by the window, working on a laptop, was the very person who had stuffed him into the locker so many years ago in high school. All the old emotions came rushing back to him at once—the humiliation, the pain, the loneliness. He choked back a tear. "My God," he thought, "I'm the most important man in the world. People do whatever I say. Why am I getting so emotional? Am I ever going to get over this?" He pulled himself together enough to sneak behind the man and look over his shoulder. There, on the laptop screen, he could see that the man was running his program. He turned away and smiled, and then his smile broke into a broad laugh that reverberated around the room. People looked up from their devices and stared. He brushed a tear out of the corner of his eye. "It looks like I won after all," he thought to himself, and triumphantly got in line to order his latte.

Article reprinted from *Christian Summer Camp Magazine*. Vol. 37. September 2020.

VR CHRISTIAN CAMPS: THE LATEST IN PROGRESSIVE CHRISTIAN CAMPING
PART I OF III

T*he times they are a-changing*, said Bob Dylan back in the sixties. Nowhere is that more evident than in the world of summer camps, especially the Christian ones. Traditionally, summer camp was seen as a place where kids could get away from it all, go back to nature, develop lasting friendships and grow in the process. Christian camps added the spiritual element as well. Kids could go up the mountain and have a genuine 'God' experience, far from the distractions down in the valley below.

Technology was seen as the enemy to all of this. You weren't likely to experience the awe of a quiet mountain vista if your cell phone was always going off in your pocket. You weren't going to hear God speaking to you while you were trying to get the high score on your video game. So, typically, Camp Directors banned cell phones and all electronics from camp. This led to a predictable backlash—both from the campers who loved their games and phones, and from the helicopter parents who wanted to monitor their child's every activity. Camp Directors were getting worn out by the endless conflict.

Happily, there now seems to be an answer. It's from a new company called *VR Christian Camps*, the brainchild of its founder and CEO, summer camp guru William Graham Muir. It operates on a basic premise—since VR (Virtual Reality) can now mimic the natural milieu so accurately, right down to the wind on your face and the smell of pine trees in the air (not to mention the taste of lousy camp food), *there's no reason to have to actually send your kids to camp at all anymore.* And since campers will be in a perfectly safe virtual environment, their parents won't have to worry about whether Little Johnny will be stung by a bee or not get along with the other boys in his cabin. It's a win-win situation for all. As the motto of the camp says:

VR Christian Camps
All the benefits of the Christian Summer Camp experience—with none of the drawbacks!

We were fortunate enough to be able to sit down with William Graham Muir, directly, face to face, in person, via Skype:

Hi, William. Thanks for being with us today. How did you first get the idea for VR Christian Camps?

"Thanks. First of all, you can call me 'Billy'. I'd like to start out by emphasizing that we're doing this for the glory of the Lord. Through the years, Christian camps have provided people with some of the most powerful spiritual experiences they will ever have. By getting away from it all, kids could really "tune in" to God. But the whole thing was turning to chaos. People were fighting with each other. Kids were refusing to go to camp if they couldn't bring their phones. In Southern California, some churches just gave up and starting sending their kids to 'camp' in luxury condos with waterslides across the street. It all got so *tacky*. I mean, what was next? Pink flamingos on the church lawn? Pastors wearing make-up? So when the Lord put Romans 12:18 on my heart, '*If it is possible, as far as it depends on you, live at peace with everyone*', and with me being a TFG, I felt I had my calling."

TFG?

"Techie For God (laughs). Anyway, since the technological world was already intruding on the summer camp experience, we decided we'd be better off joining it rather than fighting it—but in a way that honored both nature *and* Christ."

Tell us about some of the problems you were seeing. Any interesting stories?

"Actually, the kids weren't the main problem. It was the parents. They were the real tech addicts.

Parents were sending kids to camp with three phones. A kid would turn one in to us, keep one in his suitcase, and hide a burner phone out in the woods in case the second one got confiscated. When we started discovering phones buried beneath trees and hidden under rocks, we knew something was up. Then one day when an eight year-old girl came to us crying, saying that a squirrel had taken her phone up in a tree and was taunting her with it, the whole scheme unraveled."

What happened when you called her parents?

"We talked to her mom, who threated to sue us if we didn't do something about the animals bullying the children. She wasn't upset about the phone—she figured her daughter still had the extra one in her cabin. She didn't know that we had already confiscated that one too."

But hadn't camps already tried to pacify the parents via video?

"Sure. For years we had been sending kids home at the end of the week with a DVD of camp highlights. We knew kids were always exhausted by the end of camp, and the last thing they wanted to do when they got home was have a big conversation with anyone. Parents, meanwhile, were eager to hear all about their child's week, and wanted to *talk*. The DVD solved the problem. Kids could just hand it to their parents when they walked through the door and go straight to bed. The kids were able to share their camp experience with their parents, *without ever having to talk to them at all.* It was great! But it wasn't enough."

Why?

"As time went on, parents became less and less patient. They didn't want to have to wait all the way until Saturday to find out how camp had gone. We also got complaints that their kid hadn't been in the video enough times, so it wasn't worth their $19.95."

So what was the next step?

"The next thing that we tried was a website. We uploaded photos of events from camp each day. Parents could log-in and see pictures of their kids doing various camp activities, and make sure that they were okay."

That sounds like a great idea—using technology to solve a problem created by technology. What happened?

"We started noticing that the same few kids seemed to be in all the pictures. It turns out, some parents were telling their children to get in as many photos as possible so they could keep an eye on them. These kids then started picking their camp activities accordingly. For example, if Little Susie knew when the photographers were going to be at the ceramics station, she'd make sure to be doing ceramics at that time, in order to make her parents happy.

Also, some parents were having their children flash signs in the daily photos to communicate how they were feeling. Eyebrows up—the child was feeling well. Eyebrows in a scowl—the child was sick. Then we'd get e-mails or calls from the parents of the scowling kids asking why their counselors weren't doing anything about it. We also started getting e-mails from the *other* parents wondering why so many scowling kids seemed to be having such a bad time at camp!"

The counselors must have been frustrated.

"They were. Especially because they knew that the parents had selected them in the first place. We had staff members post their résumés and photos online during registration, so that parents could choose the counselor who was the best 'fit' for their child."

You really went out of your way. So what was the next step?

"Real time video. 24/7 coverage from various strategic

locations around the camp. That way Little Johnny could be freed up to do any activity he wanted during the day and still be under parental supervision."

Wow! Giving the children complete autonomy while keeping them under constant surveillance. That's brilliant! Did it work?

"Nope. It still wasn't enough. One parent complained, for example, that there wasn't an underwater camera in the pool area. How could he know if his son, who was 13 and still wearing floaties, was holding his breath long enough? What if he hadn't put the floaties on right? What if he *sank*? We tried to explain that we didn't want to show kids underwater in their bathing suits because of online predators and so forth, but he still threated to sue if we didn't add another camera."

What happened?

"We stood firm. We figured we'd win this one in court. We refunded the man his money and he took his son home. *I still don't think he can swim.*"

So how did all of this lead to VR Christian Camps?

I overheard a conversation between a mother and daughter in the parking lot one pick-up day. The mom was crying, saying, 'I was so afraid. I thought that maybe you'd forget all about me since I wasn't texting you every night.' When I heard that, the little light bulb went off above my head—Virtual Reality Camps! What if we could provide kids with a real camp experience—all while they were sitting in the Fellowship Hall at church wearing VR goggles? Their parents, meanwhile, could be right next door the whole time having a Bible study. Or whatever. We started working on it right away."

Next month in Part II: The second part of the interview, including details about this exciting new direction in Christian summer camps.

Article reprinted from *Christian Summer Camp Magazine.* Vol. 38. October 2020.

VR CHRISTIAN CAMPS: THE LATEST IN PROGRESSIVE CHRISTIAN CAMPING
PART II OF III

In this month's issue, we continue our conversation with William Graham Muir, founder and CEO of *VR Christian Camps.* Last month, we learned how the tech war between campers, parents and Camp Directors became Muir's inspiration for this exciting new direction in summer camps. Here in part two, he discusses some of the ins and outs of the business end of things. Part 3 will illuminate the day-to-day *VR Christian Camps* experience.

Once again, we are sitting down with Muir, directly, face to face, in person, via Skype:

I was wondering if you could give us a brief overview of how VR Christian Camps work.

"I'd be happy to. Here's the basic scenario from a typical church: parents drop their kids off in the church parking lot early on Sunday morning, like always. Instead of loading them into vans, or God forbid, those terrible church buses, they are led into the Fellowship Hall. Each kid has his or her own assigned seat, which was individually pre-calibrated the previous Thursday night. The seat itself contains the electrodes and sensors necessary to implement the VR technology. Our technicians come around and make any last second adjustments. The kids then put on their goggles and are immediately ushered into the VR environment."

It sounds like all this might get really expensive, especially for smaller churches.

"That's true. But we know that most kids already own a set of VR goggles—the ones they use for school. Our church got really proactive about this. We used to give Bibles to the children at the beginning of third grade, with their names embossed on them in gold. Since nobody reads anymore, we came to the realization that we were not being faithful stewards of the resources God had provided for us. So we started giving out VR goggles instead. Each set had the kid's name on it, plus an inscription of John 3:16. That way, kids could wear them at school too, and be a witness to all of their friends."

What about the cost of the seats themselves?

"While some of the big mega-churches are able to buy a set, most churches just rent them. We were fortunate to get a large influx of cash from a venture capitalist in Silicon Valley who became a Christian at Hume Lake Camp back in the 80s, so we are able to subsidize the cost to a significant degree. Most small churches end up not paying anything at all."

Is this a sustainable business model? It sounds like you're just spending down your cash reserves.

"You're right—but I'm old. I figured it was time to start giving back some of what the Lord had blessed me with.

Besides, the technology will eventually catch up to my vision. I predict that in the next twenty years, as the costs of VR technology come down, most churches, at least in America, will be going to VR for pretty much *all* of their ministries."

All of their ministries?

"Yes. For example, we're already hard at work on *VR Sunday School*, *VR Vacation Bible School*, and even *VR Church Potluck Supper*."

Wow! It's going to be really exciting to see what the Lord is going to do with all of this.

"We think so. But there are still some detractors—mostly among people my age. They're still stuck with the Neo-Luddite idea that people need to actually get together face to face in order to have true communion with each other."

That leads right in to my next question. Why bother to have kids meet in the Fellowship Hall at all? Why not just have kids plug in via a port at home?

"First of all, poor kids don't necessarily have ports in their homes. And they're not going to be able to access VR Camps effectively from, say, the public library. Can you imagine some poor harried librarian trying to chase these kids out at closing time? So we wanted to make sure to bridge that economic gap.

It's also a quality control issue. We feel very strongly that it is our responsibility to give an *authentic* presentation of the gospel, and our VR technology is, we like to say, '*more real than real.*' If a camper perceives that there is something not quite right about the shape of a certain cloud, for instance, the VR illusion will be broken, and he'll start critiquing everything: *the Sloppy Joes don't taste right*, *the water in the lake is too cold*, etc. At that point the possibility of the kid coming to Christ is almost always lost, according to our research. So we prefer to keep everything in-house."

Any other reasons?

"Actually, the main reason is biblical. Hebrews 10:25 admonishes us to *not forsake the habit of meeting together.* And of course Jesus said that *wherever two or three are gathered in my name, there will I be also.*

So we feel like we are being obedient to the scriptures by having the kids meet together. And this is an even more exciting thought: can you imagine where this can all possibly lead? I mean, we're talking about *the power of God*, *enhanced by the power of technology*! C'mon! It's like a Holy Ghost explosion! Ka-boom!"

Have you considered marketing this to other religions as well?

"Absolutely. After all, the very idea of the religious life is being affected by technology all across the world. We are, in a sense, despite our differences, all in this together. Buddhist monks in Tibet, for example, are going *nuts* because cell phones keep going off during their meditation sessions.

Now, since we *are* a Christian organization, we're not going to directly manufacture products for anyone else. We are, however, currently in non-profit partnerships with other religious companies to develop *VR Bar Mitzvah*, and *VR Hajj: The Pilgrimage to Mecca.* Future products down the pipeline include *VR Buddhist Meditation* and *VR Zen Contemplation.* Hinduism, unfortunately, has proven to be very tricky—there are so many gods, it's almost impossible to replicate them all with current VR technology."

It sounds like there just might be hope for religion after all. Hallelujah!

"We certainly hope so."

Next Month: Part III: A day at VR Christian Camps.

Article reprinted from *Christian Summer Camp Magazine*. Vol. 39. November 2020.

VR CHRISTIAN CAMPS: THE LATEST IN PROGRESSIVE CHRISTIAN CAMPING.
PART III OF III

This month, we conclude our interview with William Graham Muir, founder and CEO of *VR Christian Camps*. So far, we have learned about the conflicts that initially led to the camp's creation, and some of the business decisions involved in its day-to-day operations. In this month's issue, we take a look at what it's like to experience a day at *VR Christian Camps*.

Once again, we are here with Mr. Muir, directly, face to face, in person, via Skype:

Hello again, Billy. So what is it that makes the VR Christian Camps experience so unique?

"I'm not sure that it is *so* unique. After all, we're really not that much different from any other camp, except for the fact that none of it is real."

What do you mean?

"What we're trying to create is an entirely artificial, yet still incredibly natural environment. As it says in Psalms, *the heavens declare the glory of God.* We want kids to *see* that, to *touch* that, to *feel* that glory. Virtually, of course. When they walk in the first day, we want to wow them. So far, it seems to be working. Again and again, they tell us how much better our simulations are than the ones they have in their video games back at home."

So you've been able to mimic nature exactly?

"No—that would be impossible. It's way too complicated. We do catch one break, though. Since kids never go outside anymore, they don't *really* know what nature is supposed to feel like."

Are there any other key features of the camp?

"Lots. Our research tells us that one of the problems with camp in the old days was homesickness. You'd have kids crying in their bunks at night missing their parents. Counselors didn't know what to do. But now, the parents are right next door. If a camper gets homesick, all she has to do is take her goggles off, and mom will be right there. Remember, the parents are there all week too."

I love it! Full adult participation in children's activities. There's just not enough of that in our society today.

"Let me tell you a funny story. We had a little boy get homesick the first night of one of our camps. He took his goggles off and went to look for his mom. When he found her, she hardly recognized him. They were so used to communicating with machines as intermediaries, she wasn't quite sure what to do with him in real life. 'Get back in there,' she said, 'so I can give you a virtual hug!'"

That's really funny! Do you ever have adults try to sneak in as campers?

"Yes, and that's even more hilarious. They always try to hide their identities by coming to camp as avatars, but end up giving themselves away by choosing outdated ones, like Super Mario or Papa Smurf."

Can kids come as avatars too?

"Sure. Our term for this is *layering.* It's especially great for shy kids who need safe spaces. The VR environment itself provides the first layer of distance between them and the outside world. The avatar provides a second layer. Avatars who play roles in camp skits get a third layer of protection. If the skit is then shown on video, that's still *another* layer. And so forth and so on *ad infinitum.* Layer upon layer upon layer. The only limit is the child's imagination. One kid actually got in so deep that he got lost and never came back! We still don't know where he is."

Really?

"Yep. His church eventually asked us to remove his body from the Fellowship Hall. They needed the room for Bingo."

What did you do with him?

"We've got him in a closet in our office. He's still strapped into his seat. We're monitoring his vital signs, in case he ever shows up."

One of my favorite things about camp was always the singing. How do you handle that?

"We still sing the old goofy camp songs, of course. But because these are Christian camps, we also want to provide powerful times of praise and worship for the campers.

Rather than just having some hippie girl strumming an out-of-tune acoustic guitar singing *Kum-ba-yah*, we bring in, in holographic form, the latest bands from the top of the Christian charts. There's a rockin' concert every night, dude!"

Awe-some! That leads us right to the thing I wanted to discuss most—the spiritual aspects of the camp. Besides the singing time, what is there for campers to experience spiritually?

"The *very thing* we're most proud of. The *very thing* that we feel fully maximizes the potential of VR as a ministry tool. We call this powerful new technology *Incredible Bible Simulations.* Kids get to experience Bible stories for themselves, in true life-like form."

Can you give me an example?

"Sure. In order to teach kids about God's omniscient power, we put them on a virtual Ark with a virtual Noah. We've tried to recreate the actual experience of the Great Flood. The seas rise in a huge storm. Rocks, trees, and eventually mountains are submerged. Animals die. You can feel the force of the wind and taste the salt-water as the waves crash against the Ark. It's *loud.*"

Are the younger kids scared?

"Heck—*I'm* scared. It's *super* realistic. The worst part is when the last few people clinging to the side of the Ark let go and fall screaming to the depths below. It's a real thrill ride, that's for sure."

It sounds terrifying.

"It is, but we don't leave them hanging. Eventually, the storm calms, the clouds open, and the sun shines. Virtual Noah releases a virtual dove, and they see a virtual rainbow—God's virtual promise to all of virtual humanity. It's so *real.*

Just then, at that strategic moment, we project Romans 6:23 onto a cloud in large block letters. A deep, somber voice recites:

> FOR THE WAGES OF SIN IS DEATH, BUT THE FREE GIFT OF GOD IS ETERNAL LIFE THROUGH CHRIST JESUS OUR LORD.

The kids are so scared they can't help but accept Jesus as their personal Savior—right then and there! They've *seen* death. They've *experienced* salvation. *They love free stuff!* They're ready to commit. Who wouldn't be? And all this in just under an hour! It's sooo efficient! Ka-*boom!* Are you kidding me? Billy Graham *wishes* he could have had this technology at his disposal! Ka-*boom!* (*he pounds the table in front of him several times, then calms down*). Of course some of the kids just get seasick, and puke over the side of the Ark. Many of them throw up in real life too. That's why we always tell kids to pack an extra set of clothes for camp! Our technicians clean up the mess in the Fellowship Hall, so the kids can stay in VR and connect with the Lord."

That sounds amazing! Are there any Bible stories that you don't do?

"Well, we used to have a *David and Bathsheba* simulation, but the younger campers never chose it. It was just too *weird*, they said. They didn't get it. It seemed like the only ones using it were our computer programmers back in the offices late at night. So we scrapped it.

We also don't do simulations for any parts of the Bible that the kids think are boring. It's a pick and choose world, after all, and we feel very strongly that people only need to read the parts of the Bible that they like. Let's face it, a lot of the Bible is *pret-ty* dull, and we know for sure we'll never reach this generation of kids if they're bored. Besides, they haven't read their Bibles anyway, so they don't even know which stories are missing.

Our most popular simulations are *Noah and the Ark*, *The Battle of Jericho*, *David and Goliath*, and of course, *The Book of Revelation.* We don't simulate the Crucifixion or the Resurrection."

Why not? Did you think it might be sacrilegious?

"No. It's just that we could never get everyone at the office to agree on what Jesus looked like."

Well, it seems we're running out of time. Thank you so much for talking with us—this has been fascinating. I think that there is an incredible future ahead for Christian camping, and that VR Christian Camps will certainly play a large role in that future. We wish you all the best. Is there anything else you'd like to tell us about VR Christian Camps?

"I could tell you how much more likely kids are to confess their sins when they come to camp as avatars, or how great it is that you can go to camp at a different location every year, or about how about parents no longer have to worry about their daughters coming back with stories of kissing their camp boyfriends, or. . ."

Our Skype connection abruptly terminated here.

THE HEAVENS DECLARE THE GLORY OF GOD, 1818

John awakens to a cold, Kansas morning. Rising before dawn, he milks the cow while his sister feeds the chickens and the pigs. Ma makes breakfast. Pa goes down to the well to get water. Work is a family duty. Survival is their occupation.

After breakfast, Pa pulls out the family Bible. "Psalm 19," he begins. "The heavens declare the glory of God, the firmament shows his handiwork." John is struck by the words. They resonate within him as Pa reads the rest of the psalm.

The heavens declare the glory of God, the firmament shows his handiwork.

"I wonder if that's true," he thinks as he starts out to the fields. Everything is calm. The sun is just coming up over the hills. Across the fallow wheat field a soft breeze is blowing. His horse nickers quietly to him as he puts on her harness. They walk out together. The first birds have returned after the long winter. He can hear them chirping in the trees.

The heavens declare the glory of God, the firmament shows his handiwork. He says the words again to himself.

On this day, they are preparing the field for planting. He leads his horse down each furrow, and she pulls the plow behind her as she goes. The work is hard for both of them. The sun rises and is hot overhead. For much of the day he is quiet, alone with his thoughts.

He thinks about the seed that will soon be sown, and the rains that will come. Soon the new wheat will sprout. He remembers a book he saw once at the little schoolhouse. It showed drawings of the life cycle of a seed. He didn't need a book to tell him that. What he really wanted to know was *why* the seed did what it did. What *made* it grow? His teacher didn't know. The book didn't say.

He doesn't think much of school. What good are facts, if you don't know their meanings?

He thinks about the seed, and the plants, and the mystery of things.

The heavens declare the glory of God, the firmament shows his handiwork.

At noon his sister brings him his lunch in a pail. They talk softly together for a few minutes before she returns to the house. He and his horse stop to eat beneath the shade of a large elm tree. He lays down in the shade. He is just about to nod off to sleep when he is startled by a strange sound.

In the distance he can see a large black storm cloud looming on the horizon. He worries about Pa, who is out hunting. As it gets closer he sees that it is not a cloud but a massive flock of birds. With a giant whirlwind and a clambering like thunder, they sweep over him and around him. The sky grows dark with the beating of wings. His horse neighs in fear. It takes twenty minutes for the entire flock to pass by.

The heavens declare the glory of God, the firmament shows his handiwork.

He goes back to the fields to finish his work. When the long afternoon finally comes to a close, he and his horse head back in for the night. He does his evening chores and the family gathers around the table. As they eat together they take turns talking about the events of the day. Each one has something to say. He describes the flock of birds. He tries to convey the sense of awe and wonder that he felt, but the words won't come.

His father knows what he is thinking. "Come with me, all of you," he says. The family walks out into the starlit night. "Look up." Millions of stars press down upon them like a weight. From horizon to horizon, the sky is filled. They stare in wonder. It is more than enough.

Pa breaks the silence.

"The heavens declare the glory of God, the firmament shows his handiwork," he says. The quiet wind makes the words ring out like chimes. "That's what you mean, isn't it?" He looks at his son. John nods.

"I want you all to remember this," Pa's voice trembles. They stand for a long time in silence, and then go back inside.

They will never forget.

THE HEAVENS DECLARE THE GLORY OF GOD, 2018

John awakens to his ring tone on a cold, Kansas morning. Rising before dawn, he checks his in-box. Three e-mails have come in while he was sleeping. He skims a promotional e-mail from Nike.com, one from Spotify about new playlists that have been added, and one from his school about Spirit Week. The rest of the family is awake, checking their devices.

After breakfast, his father pulls out his tablet. "Psalm 19," he begins. "The heavens declare the glory of God, the. . ." John feels a vibration coming from the phone in his lap. It's a text from his best friend Steve: "can u tell Mrs. Q that I'll miss 1st period?" "Yup," he answers. His dad is still talking.

"Can you repeat that?" John asks him.

"The heavens declare the glory of God, the firmament shows his handiwork," he says.

"I wonder if that's true," he thinks as he goes out the door for school.

John is sixteen, but does not yet have his driver's license. His mom turns on talk radio in the car. On the way to school, he gets a Facebook notification from his friend Billy, tagging him and four other friends in a comment under a Shaquille

O'Neal video on TNT. The text says: "John James Jake Tyler Steve dudes check this out!! LOL." As they drive in the parking lot, he gets a text from Jenny, a girl he's interested in: "good morning :) :) see ya in 2nd per!!!"

His first period English class is reading *Macbeth*—something about three witches. His phone vibrates. It's a text, from Jake to John, Tyler, and Steve: "this class is killing me"

Steve: "duh"

John: "always"

Jake: "can't wait to get out of here"

They go on complaining about the teacher and school in general. He tries to focus on the words in the book, but they swim in front of his eyes. Thankfully, the bell rings.

John scrolls through his FB page on the way to History. The teacher is late. As they stand around the door, three of his friends look at their phones while two argue about last week's *South Park* episode. The three on their phones intermittently add to the discussion. The teacher finally arrives, and class starts. He texts Jenny: "why do we have to learn about the French Revolution anyway?" "idk," she texts back, "the french just lose LOL." They text back and forth as they try to take notes. The bell rings, and John goes out to the quad.

He and his friends stand around together during brunch. They're all on their phones, scrolling through their Instagram feeds. They tell each other which girls have posted cute photos.

Third period Geometry starts with a quiz. His teacher always posts the scores by the end of the day. His mom is addicted to School Loop and checks it constantly. If he gets less than 8/10, he knows she will be waiting for him at the door when he gets home, wondering why. After the quiz, he works on proofs for the rest of the period.

Fourth period is Band. He plays percussion. While the teacher is working with the flute section, he sends out a group text: "same spot 4 lunch?" In quick succession, three of the four guys respond with variations of "yes." Tyler never responds. For the rest of the period they work on music for the upcoming recital.

While they're messing around in the quad at lunch, three of his friends record Snapchats and post them to their Snap Story. John watches other Snap Stories that have been posted by his friends.

Jenny texts him: "excited for 5th period later?"

John responds: "nope, never am. How are u excited about it?"

Jenny: "idk, I just kinda like Chem! :P"

John: "nerd ;P"

John realizes that even though he really likes Jenny, he has not actually spoken to her all day. He checks his Twitter feed, and posts his own tweet: "another day, another boredom." Two of his buddies *Favorite* it, and a friend he hasn't seen all day reposts the Tweet. John does a quick one-minute check of his FB. One of his friends posted an article about Kobe Bryant's legacy, so he follows the link and begins reading. He doesn't finish it but tells his friends about it anyway.

In Chemistry, he receives a text from a different girl: "great snap story, loved it! :) :) :)" He doesn't respond.

Steve sends a group message: "madden after school?"

John: "can't, have practice"

He peruses Instagram behind his Chemistry book. He likes one of Jenny's photos that she posted at lunch.

During sixth period, he's a TA for his English teacher. Since she doesn't have any work for him that day, he spends the period playing games on his phone. After class, he puts on his headphones and listens to the Pump-Up Playlist he's created for himself as he walks to the gym for basketball practice. He scrolls through Spotify to see if he wants to listen to any of the pre-made playlists, and then decides against it. He does one more quick check of Instagram before practice begins. He remembers that he has a picture from the night before with his buddies that he wants to post.

After practice, he picks up his phone to check his notifications. He's been tagged in two comments on Facebook about two other NBA-related videos that have been posted.

He texts the group: "u guys still playing"

Steve: "nah we finished up a little while ago"

His friends who were playing XBOX send messages back and forth trash-talking about their earlier performance, but John doesn't respond.

Jenny texts: "what r u up to tonite?"

John: "nothing, just HW. Hbu?"

Jenny: "same, not much going on!!"

When he gets home his family gathers around the table for dinner. As they eat, he remembers to post the photo on Instagram. His sister is watching a cute kitten video on her iPad while his mother responds to a text on her phone. His dad is exasperated. "Come with me, all of you," he says. The family walks out into the night. They can hear the faint hum of the freeway a mile from their house. "Look up." A few stars manage to flicker through the glare of the city lights.

"The heavens declare the glory of God, the firmament shows his handiwork, " his dad says. "I want you all to remember this."

"That's great, Dad. Too bad we can't give God a 'like' for that one!" John laughs. He hurries back in the house. He really does like the verse, but he needs to check his phone. He has already received thirty IG likes. "Not bad," he thinks.

He sends a group text: "what are we doing this weekend boys?" He gets responses from all but Tyler, proposing ideas. Eight messages go back and forth. Tyler finally responds once the plans are made: "sounds good to me."

John goes upstairs to do his homework. He takes breaks from *Macbeth* every ten minutes to check his "like" count. He's up to sixty by 8:00. He wants to hit at least one hundred by the end of the night. He checks his e-mail. There's another promotional e-mail from Nike. Then another one from Adidas. John skims the e-mails and then reads a few more lines from *Macbeth*.

At 8:30 he takes a break to watch the Warriors play the Lakers on TNT. It's a big game and he's really excited. His friends are all watching too. They text each other each time LeBron James makes a no-look pass or Steph Curry hits a long range three. After the game he watches the post game show, and then tunes in to Sportscenter on ESPN for half an hour. He finally gets back to his homework at 11:30 and finishes his geometry problems by midnight.

He grabs his laptop and gets on Reddit. He spends an hour reading and responding to different message boards under his username "truehooper25". He receives a FB message from Jenny, dropping a hint that she wants to see a movie this weekend. John: "alright, u want to go on Saturday?" Jenny: "sure, sounds good to me! :)"

While getting ready for bed, he scrolls through Instagram one last time. He's only up to ninety-three likes. He sees that someone he knew in middle school has posted a photo, likes it, and then visits his IG page, scrolling through all his friend's photos to see what he's been up to. He gets one last text from Jenny: "g'night! :)" John: "see you tomorrow." He finally falls asleep around 2:00, the glow from his phone gently lighting the pillow by his head.

FUN!
TECHIE BIBLE STORIES!

THE RICH YOUNG RULER: REVISITED

As Jesus started on his way, a man ran up to him and fell on his knees before him. "Good teacher," he asked, "what must I do to inherit eternal life?"

"Why do you call me good?" Jesus answered. "No one is good—except God alone. You know the commandments: You shall not murder, you shall not commit adultery, you shall not steal, you shall not give false testimony, you shall not defraud. Honor your father and mother."

"Teacher, he declared, "I have kept all these since I was a boy."

Jesus looked at him and loved him. "One thing you lack," he said. "Go sell all your devices: your desktop, your laptop, your phone, your tablet, your game consoles, headphones, iPods and screens of all kinds. Close your social media accounts. Give all the money to the poor. Then come, follow me."

At this the man's face fell. He went away very sad, because he just couldn't live without his devices.

see Matthew 19:16-22

WALKING ON THE WATER: REVISITED

Then Jesus told the disciples to get into the boat and go on ahead to the other side of the lake. After sending them away, he went up on a hillside by himself to pray. By the time evening came, the boat was far out in the lake, tossed about by the wind and waves.

Shortly before dawn, Jesus went out to them, walking on the water. When they saw him they were terrified. "It's a ghost!" they said and cried out in fear.

Jesus spoke to them at once. "Take courage! It is I. Do not be afraid."

Peter spoke up, "Lord, what should I do?"

"Come to me," said Jesus.

"But Lord, you haven't seen this boat. It's very solidly built. I think it'll keep us safe."

"Go ahead, step out on the water," Jesus answered.

"What *I'm* saying," said Peter, "is that we should avail ourselves of all the technological tools at our disposal. This *really* is a well-designed boat."

"You don't need to be afraid," said Jesus.

"I'll tell you what," said Peter. "How about if we stay in the boat? Then you can come up behind us and push us to the shore."

Jesus stood, patiently watching them. Just then an even bigger boat came by. Its crew stood on the deck, beckoning the disciples to come on board. They talked among themselves for a moment, and then climbed onto the larger vessel. It started to drift away from Jesus.

"Look at this *new* boat!" Peter cried back across the water. "If you'll come and push us now, Jesus, there's no telling what we might accomplish together!"

The disciples could see that Jesus was saying something, but they could no longer hear him in the wind. Soon, he was lost in the fog.

Eventually the waves died down. "Truly, this boat is amazing!" they exclaimed.

"What about Jesus?" one of them asked.

"I haven't lost faith," said Peter, "I'm sure we'll hear from him eventually."

see Matthew 14:22-33

THE WIDOW'S MITE: REVISITED

Jesus looked around and saw rich tech titans, business leaders and financial wizards making multi-million dollar donations to charity. He also noticed a very poor widow donating a five-dollar bill. He said, "I tell you that this poor widow gave more than all the others, for the others offered their gifts out of their wealth; but she, poor as she is, gave all she had to live on."

see Luke 21:1-4

AT THE CRUCIFIXION: REVISITED

Two robbers were crucified with him, one on his right, and one on his left. Those who passed by hurled insults at him, shaking their heads and saying, "You who are going to destroy the temple and build it up in three days, save yourself! Come down from the cross, if you are the Son of God!"

We follow their conversation as they stand off to the side:

"Do you think we were too hard on him?"

"Probably."

"I'm just so disappointed in how things have turned out."

"Seriously. Just look! His ministry is literally *dying*—right before our very eyes."

"He should have listened to us."

"I know. We really could have helped him. I mean, his congregation never even grew past twelve, and people are now saying he's even lost one of those!"

"I tried to tell him about my *5-Step Plan for Ministry Growth: How to Take Your Congregation to 25, 50, 100 and Beyond,* but he just smiled at me. He didn't even answer. I thought it was condescending, frankly."

"I got that smile a lot, too. You know, he never figured out how to use analytics. I tried to show him how he could measure his ministry's effectiveness, but he wasn't interested. It's not really that complicated."

"You're right. He never even did any *research.* Remember that day when he fed the 5,000? I talked to him afterwards. Wouldn't it have been more effective, I asked, if the disciples had given out registration cards to each person at the same time they were passing out the loaves and fishes? That way, at the very least, he'd have their names and addresses to use for ministry follow-up afterwards."

"It's no wonder his ministry is dying! Doesn't he understand that we're living in desperate times? Doesn't he realize that you can't just keep passing up these types of opportunities if you expect your ministry to grow? The world is counting on us to spread the good news—it's critical."

"But that would have involved planning. He was always way too spontaneous. He would interrupt what he was doing just to help someone, even if all they did was something stupid, like touch his cloak."

"So on top of everything else, you're saying that he had lousy time management skills?"

"Terrible. He was always so inefficient."

"I told him over and over that his ministry would not be effective unless he had a *strategy.* I told him that he should embrace every new tool that came down the line, even something as simple as registration cards."

"Not only that, but I think part of the reason his ministry is failing is because he just wasn't friendly enough—especially to the strategic people who could have helped him most. All of that talk about snakes and vipers and white washed tombs. Once, he even got physically violent—in the temple! He was always a little *intense."*

"He definitely missed the lesson about the connection between his personal presence and his ministry's effectiveness."

"He had a presence alright. It was just so, *unprofessional.* Like his sermons. Always droning on and on, *you have heard that it was said, but now I tell you*, blah, blah, blah. So arrogant, and half the time I didn't even know what he was talking about."

"And he didn't use any visual aids. Nobody, I mean nobody, tries to teach without visual aids anymore. Especially if you're trying to reach the younger generation. I read a study that said that they can't even learn without them anymore."

"Right. This is a new millennium, after all. The fact is, he did not use the *very* instruments which the kids of today need in order to communicate and interact."

"And what about those parables? I bet people will be trying to figure those out for the next couple thousand years! People should just come right out and say what they have to say, is what I always say."

"So, do you think he was the Son of God?"

"I used to. But he was never willing to use all the powerful tools at his disposal—and isn't that necessary in this day and age if someone wants to carry on an effective ministry? I mean, the times they are a changing, Jesus, whether we want them to or not. Get with the program!"

"I would have been happy to donate my time as a consultant."

"Me too. But I don't think he would have listened."

Just then, Jesus looked over at them with pity.

Father forgive them, for they know not what they do.

"What did he say? I couldn't hear him. This is all just *so* sad. Just think about it—with all his power he could have broadcast his message around the globe—instantaneously."

"Shhh, he's saying something else. Let's listen."

My God, my God. Why have you forsaken me?

"Is he serious? He should have asked us that. We could have given him the answer."

see Matthew 27:38-46

THE TEMPTATION IN THE WILDERNESS: REVISITED

The techie messiah was led by the Spirit into the desert to be tempted by the devil. After fasting forty days and nights, he was hungry. The tempter came to him and said, "If you are so powerful, tell these stones to turn into bread."

Realizing that he was being tested, he panicked. "Hold on just a sec!" he said. "Let me check my phone."

He typed in the words *stone* and *bread*, clicked on a page and then broke into a big smile. "You can't fool me, devil!" he answered. "For lo, it is written:

> *A good way to make* ***bread*** *is to place it on a baking* ***stone*** *to keep it from burning. Some cooks recommend sprinkling corn meal on the* ***stone*** *first to prevent the* ***bread*** *from sticking."*

Then the devil took him to the holy city and had him stand on the highest point of the temple. "Since you claim to be so godlike," he said, "throw yourself down. For it is written that the angels will come and bear you up."

The techie messiah quickly typed in *throw, down, angel and bear,* delighting all the while in his own resourcefulness. "Well *i-ma-gine* that!" he said a few seconds later. "It just so happens that it is also written:

> *Today on Amazon all of our* ***down throw*** *pillows are on sale for only $14.99. The* ***angel*** *and* ***bear*** *pillows are our most popular!"*

Again, the devil took him to a very high mountain and showed him all the kingdoms of the world and their splendor. "All this I will give you," he said, "if you will bow down and worship me."

This time, his thumbs practically flew across the keypad, typing in *splendor*, *bow* and *worship*. "Away from me, Satan!" he crowed in triumph. "For it is written:

> *You can spend an exotic three weeks this winter* ***worshipping*** *the sun in the* ***bow*** *of a Carnival* ***Splendor*** *cruise ship!"*

The devil went away greatly astonished. For the first time in millennia, things seemed to be trending his way.

see Matthew 4:1-11

THE STILL SMALL VOICE: REVISITED

Then Elijah went out and stood before God on the top of the mountain. The Lord passed by and showed him a furious twitterstorm that mesmerized the media and dominated the news cycle—but the Lord was not in the twitterstorm. The twitterstorm stopped and Elijah heard the clanging bell from a NYSE technology IPO—but the Lord was not in the bell. After the bell, the blogosphere erupted with rumors of a disruptive new algorithm that could mine massive data sets—but the Lord was not in the rumors.

Finally, there came a still, small voice. It was the voice of God.

Elijah recognized it, and realized that God had been speaking to him his whole life. He had just been too distracted to listen.

see I Kings 19:11-12

WE KILLED THE HYMNS

We weren't trying to kill the hymns. We really weren't. We were just trying to loosen people up, to set them free to be more expressive. Instead of having their noses buried in dusty hymnals, their hands and faces would be turned upward towards God. Instead of reading musical notes on a page, they would have the melodies in their hearts. We were putting the new wine into new wineskins, we told ourselves. God was really going to use us. It was cool. So were we. That's what we were thinking when we started putting the songs up on the screens. We really weren't trying to kill the hymns. But that is exactly what we did.

Had we known better, I'd like to think we might have done things differently. The truth is, we never gave it a moment's thought. We didn't perceive of the screens as being technologies, much less understand how disruptive they would be. We didn't know that technological use always has spiritual implications.

Those who were older and wiser tried to warn us, but what old age affirms, youth often ignores. The truth is, they did not always communicate their concerns rationally or calmly. We thought they were just being legalistic. Or nostalgic. Why should we pine away for the past when we were on the cutting edge of the future? They were the ones who were quenching the Spirit, not us. Besides, a lot of those old hymns were really boring.

But technological systems grow in mysterious, self-perpetuating ways. The screens, it turns out, were much more conducive to a new type of songwriting. Since fewer words could fit on a screen than on a page, the songs would have to have fewer words. Since there were no notes to follow, the melodies would have to flatten out and become simpler. As always, the medium helped dictate the message. While the continual repetition made it easier for song leaders to induce the proper spiritual mood, it made the songs themselves disposable. You can only wring so much water out of a sponge before it goes dry, a process that usually takes two to three years with even the most popular choruses. And why were we trying to use a technique to *induce* anything? That should have been left to the Spirit.

Yet we kept on insisting and tradition kept on resisting, until it eventually became easier to just create separate church services. The result was that the old would never again have to be bothered by the young (*thank God they finally turned that noise down!*), and the young would never again have to come to church without the knowledge that they had been catered to. In the name of unity, official sanction was granted to division.

Naturally, the schism grew. Separate services became separate churches, as start-ups strove to indulge every tech-savvy sensibility. Churches were no longer identifiable by theological distinction, but by technological preference. Of course this never worked for long. The cutting edge always grows dull. The hip church of today very quickly becomes the middle-aged church of tomorrow, and what looks more foolish than a mom trying to dress like her high school-aged daughter? Sooner than later, these churches too were in need of an upgrade.

Now our sanctuaries are filled with screens. They have taken on a life of their own and have come to dominate every aspect of the service. Walk into many churches today, and the pastor has been multiplied into three twenty-foot high images of him or herself. I always wonder what George Orwell would think of these enormous projected heads with their looming faces and dilating pupils. I worry that in the future there will be more pressure put on pastors to be good looking, since the close-up now seems to be a requisite part of the job. "Can you shoot me from my good side?" they will be taught to ask in their seminary classes. And when the people are not looking up at the screens in front of them, they are looking down at the screens in their laps. What happened to the faces that were supposed to be turned upward?

Meanwhile, in most churches the hymns are almost gone. We may still sing a few to pacify the elderly, but there is no realistic hope for their survival.

"Wait just a minute," you're saying. "This is ridiculous! There was nothing magical about the hymns. Besides, isn't the hymnal itself a technology? And if the songs were old and boring and no longer spoke to anyone, they deserved to die. You're the one who is being nostalgic."

You are right, partially. There was nothing magical about the hymns. Some of them *were* just plain bad. But there is something extraordinary about a song like *A Mighty Fortress is our God.*[67] For almost five hundred years, generation after

[67] *A Mighty Fortress is Our God* turns 491 years old in 2018.

generation has discovered this hymn, been spiritually moved by it, and considered it worthy of preservation. It has outlasted nations, states, rulers, and empires. It has crossed cultural boundaries, withstood every musical, social, intellectual, and theological trend, and survived an endless number of wars. I would humbly suggest that if there is a generation that comes upon this hymn and finds it boring, the problem is not with the hymn.

And yes, the hymnal was a technology too. It had its own set of drawbacks. Because it was a book, and a book is finite, it seemed to suggest that all the good songs had already been written—that *these* songs, and these songs only, had the ecclesiastical seal of approval. This was not exactly encouraging to young aspiring songwriters and composers. The hymnals could also be bulky and inefficient, and yes, people often kept their noses down in them when singing.

But there was something very moving about leafing through the pages, looking at the names and composition dates, and imagining a person, maybe from a completely different culture, standing in a church on a Sunday a hundred years earlier and singing the same song I was singing today. There was also a musical education to be had from learning to read the notes and time signatures and trying to pick out your part. And so many of the hymns are so *great.* You can wring the sponge out a million times and *Amazing Grace* and *Be Thou My Vision* will never run dry. All of this is being lost.

But we never even thought about it. To us, it seemed like all the styles would be able to co-exist. We grew up singing hymns from the hymnal and youth group songs from the youth songbook, and for a treat our pastor Bill Simpkins would sometimes bring out his old guitar and teach us simple choruses during the evening service. It was all good.

But we couldn't leave things well enough alone, technologically. We just had to push it. We are left with a situation that is much like trying to own a horse today. Sure, it's theoretically possible that I could ride a horse to work, but the city streets are not set up for it, and there'd be no place to park once I got there. And where would the horse live? I'd never be able to get a city permit to keep it in my backyard. New technological systems eventually eliminate the possibility of using the old ones, except as a novelty. The same is true for the hymns. There will be no going back.

In the end, the real problem was never the music anyway. It was our pride—the same pride of self-sufficiency that always accompanies technological innovation; the same pride that leads us to not think through the potential consequences of our actions; the same pride that makes us believe that God sure could use a hand—from us. So maybe it's time to bring the old hymnals out of the storage closet, dust them off and give them another try. We'd have to show the children how to hold them and use them. At the very least, we'd be sending the message that we have no more interest in being cool, and that if we are to accomplish anything, it'll have to be because God is working through us.

We'd better hurry, though. It won't be long before someone develops a worship song generating app and people start pining away for the good old days when at least humans still wrote the songs.

POSTSCRIPT

THE ICARUS GENERATION

One of the Greek myths that people remember most from their school days is the story of Icarus, the youth with the wax wings who plunges to his death after flying too close to the sun. What fewer people remember is that it was his father Daedalus, the great architect and inventor, who got him into trouble in the first place.

Here's the story: Daedalus was the creator of the famous Labyrinth on the island of Crete, which he designed to constrain the terrible monster called the Minotaur. With its tangled turns and unpredictable pathways, once a person went inside there was no way to escape. Every nine years, King Minos demanded that seven youths and seven maidens from Athens be sent to Crete as tributes. They were then placed in the Labyrinth and the people would watch from on high as the Minotaur tracked them down and slaughtered them one by one.

One year, Ariadne, a young Minoan princess, fell in love with Theseus, the Athenian prince who had volunteered to come as a tribute. She vowed to do whatever she could to save him, so she went to Daedalus for help. He gave her a ball of thread that she then passed along to Theseus. By fastening it to the door and unraveling it as he went along, he was able to navigate the maze, kill the monster and retrace his steps. He and Ariadne escaped, along with all the other Athenians.

King Minos was furious. He knew that Daedalus must have helped them escape, so he had him thrown into the Labyrinth, along with his son Icarus. The Labyrinth was so well constructed that even its creator could not find the way out without any help, but Daedalus did not panic. He fashioned two pairs of wings, one for Icarus and one for himself, binding each together with wax. Just before he and his son took flight, he warned Icarus to be careful to keep a steady, middle course over the sea.

As the two started out, however, the exhilaration of flight proved to be too much for the boy. He soared higher and higher, paying no attention to his father's warnings. The wax began to melt. His wings fell apart and Icarus plummeted to the sea, calling out his father's name as he went. Daedalus watched, heartbroken, but could do nothing to save him. As the waves closed over his son, he cursed the invention he had made.[68]

In school, this story is usually taught as a lesson for young people: "Be careful, don't fly too close to the sun," students are told, and then the discussion centers around various things that this could possibly mean. The story becomes far more interesting, however, when we assume that the target audience is adults. Like so many other ancient myths, it then comes roaring back to life in new and interesting ways.

The Greek word for Daedalus is *Daidalos*, which means, "cunningly wrought". He is the main figure in Greek mythology that represents what we today would call the

---

[68] see Ovid and Hamilton, Edith.

technological impulse. His words and actions reveal what the Greeks understood about the subject, much in the same way that Ares represented war and Athena wisdom. Once again, the ancients seem to have known much more than we ever give them credit for.

Consider the following in regard to the technology of the Labyrinth:

- It is created by an adult, but ends up enslaving a child as well.
- The problems it causes are, well, *labyrinthine.*
- It was created for one purpose but ends up serving others—in ways that no one could ever have anticipated.
- Not even its creator is totally sure how it works.
- Not even its creator can escape from its effects.
- The best way to escape from it is to retrace one's steps back to the very start.
- Retracing one's steps is only possible if there is an exit plan in place from the beginning.

In regard to the wings:

- They are a technological solution to a problem caused by an earlier technology.
- The technology that ends up destroying the child is given to him by his parent.
- In the end, it makes no difference that the parent's intentions were good.
- The child is so delighted by the feeling the new technology gives him that he isn't careful when using it.

- The child doesn't pay attention to his parent's warnings.
- The technology has dangerous design flaws that the young user does not fully understand until it is too late.
- The parent is powerless to stop his child's fall.
- The adult survives, but the child perishes.

The insight here is no accident. The ancient Greeks may not have been as "advanced" as we are, but they were wise enough to perceive that technology works according to a basic set of principles and emerges from a mindset that goes back as far as we can remember.

Young people today are the Icarus generation. They are trapped in a technological maze that was created by their forebears, and have been handed a set of wings with very few instructions on how to use them. We can hardly blame them when they then choose to fly too close to the sun. We created the Labyrinth and the wings, after all, and we don't even fully understand how they work. How can we expect them to?

Sadly, this is not the first time children have been mistreated in the name of technological progress. During the Industrial Revolution, for example, children were stuffed down chimneys and forced to work long hours in dark factories while being chained to machines. We tell our youth that they are free to fly, but in many ways today's chains are even worse than those of the past. In the factories, they restricted bodily movement. Today, they are robbing young people of their very ability to think.

The chains also have spiritual implications. It is during our youth that we are the most spiritually open—to new ideas, new people, to the arts, to nature and to God. What a terrible time to be tethered to a device!

The apostle Paul puts it this way:

> *"In the past you did not know God, so you were slaves of beings who are not gods. Now that God knows you—how is it that you want to turn back to these weak and pitiful ruling spirits?"*
>
> *Galatians 4:8-9*

How is it, indeed? And how is it that we teach (and sometimes even *demand*) that our children turn to the false gods as well? It is enough to make one weep.

All analogies eventually break down at some point, and this one is no exception. For us, there is one key character from the story that is missing: King Minos. We have not been sentenced to die in the Labyrinth by some evil tyrant. Unlike Daedalus and Icarus, we can choose to just walk right out. The ball of thread has been provided by the Spirit, and the pathway to our freedom consists of communion with nature, each other and God. It is only when we retrace our steps and take that path that we can begin to live our lives the way they were meant to be lived. The good gifts of God are free, and it is only through them that we will finally find the desires of our hearts and have everlasting rest for our souls. May we all be fortunate enough to find that rest.

ACKNOWLEDGMENTS

One of the most enjoyable aspects of writing this book was sharing the process with so many other people. Each made significant contributions to the book for which I am eternally grateful: to David Gill, for the encouragement and professional advice; to Brian Quick, for the many discussions that helped me clarify my thinking on various issues; to Julie Anderson, the librarian mentioned in the *Freedom* chapter; to Paul Vaughn, both for the editorial assistance and for living out the ideals of this book far better than anyone else I know; to Jeff Reed, for ongoing poetic excellence, and for rescuing the text from several potentially fatal errors; to Corey, for help with writing clarity and authentic text-speak; to Kelly, for keeping me accountable for the book's overall tone; and to Martha, my first and best editor, for supporting and encouraging me daily through the entire process. Finally, a special thanks goes out to the unknown clerk at Robert's Bookstore in San Jose, who one day thirty years ago put an old copy of *Amusing Ourselves to Death* on the paperback rack in the front window, just for me.

ABOUT THE AUTHOR

C.M. Collins is not a minister, a techie, or a theologian. He is, however, an avid reader and student of human nature. Like so many others, he struggles on a daily basis with both his faith and his relationship to technology. He and his wife Martha live in California, and have two adult children. He does not own a cell phone.

SOURCES

Blake, William. "The Marriage of Heaven and Hell" in *English Romantic Writers.* New York: Harcourt Brace Jovanovich, 1967.

Bradbury, Ray. *Fahrenheit 451.* New York: Simon & Schuster, 1951.

Brand, Matthias, et al. "Prefrontal control and Internet addiction: a theoretical model and review of neuropsychological and neuroimaging findings". *Frontiers in Human Neuroscience,* 27 May 2014.

Carr, Nicholas. *The Shallows.* New York: W.W. Norton & Co., 2011.

Cole-Turner, Ronald Ed. *Transhumanism and Transcendence.* Georgetown University Press, 2011.

Dickinson, Emily. "Tell all the truth, but tell it slant—" in *The Collected Poems of Emily Dickinson.* Barnes and Noble Classics, 2003.

Doherty, Brian. *This is Burning Man.* Dallas: Ben Bella Books, 2006.

Dovey, Dana. "Shocking Number of Young People Report Painful and Unsatisfying Sex Lives." MedicalDaily.com, 14 August 2017.

Ellul, Jacques. *The Ethics of Freedom.* Grand Rapids: Wm.B. Eerdmans, 1976.

—*The Meaning of the City.* Grand Rapids: Wm.B. Eerdmans, 1993.

—*The New Demons.* New York: The Seabury Press, 1975.

—*The Technological Bluff.* Grand Rapids: Wm.B.Eerdman's, 1990.

—*The Technological Society.* New York: Vintage Books, 1964.

Eskridge, Larry. *God's Forever Family.* New York: Oxford University Press, 2013.

Frost, Robert. "Stopping by Woods on a Snowy Evening". *The New Republic.* 7 March 1923.

Grossman, Lev. "2045: The Year Man Becomes Immortal." *Time.* 10 February 2011.

Hamilton, Edith. *Mythology.* New York: Back Bay Books, 1942.

Havelok, Eric A. *The Muse Learns to Write.* New Haven: Yale University Press, 1986.

Hawkins, Nevel A. *The Selling Process: A Handbook of Salesmanship Principles,* 1920.

Harris, Michael. *The End of Absence.* New York: Penguin, 2014.

Hayworth, Kenneth. "Mind Uploading". *Skeptic Magazine.* vol. 21 no. 2, 2016.

Huxley, Aldous. *Brave New World.* New York: Harper & Brothers, 1932.

Isaacson, Walter. *Steve Jobs.* New York: Simon & Schuster, 2011.

Johnson, Skip R. *Codependency and Codependent Relationships.* BPDMFamily.com, 13 July 2014.

Kassan, Peter. "Uploading Your Mind Does Not Compute". *Skeptic Magazine.* vol. 21 no. 2, 2016.

Keynes, John Maynard. "Economic Possibilities for Our Grandchildren." *Essays in Persuasion.* New York: W.W. Norton & Co, 1963.

Kuhn, Robert Lawrence. "Virtual Immortality." *Skeptic Magazine.* vol. 21 no. 2, 2016.

Kurzweil, Ray. *The Age of Spiritual Machines.* New York: Viking Adult, 1999.

—*The Singularity is Near*. New York: The Viking Press, 2005.

Lancer, Darlene, *Codependency for Dummies*. New Jersey: John Wiley & Sons, Inc. 2012.

Levy, David. *Love and Sex With Robots: The Evolution of Human-Robot Relationships*. New York: Harper Perennial, 2008.

Lewis, C.S. *Mere Christianity*. London: Macmillan, 1952.

—*That Hideous Strength*. New York: Macmillan. 1946.

Marano, Hara Estroff. "Love Interruptus." *Psychology Today*. August 2016.

Mayor, Tracy. *Asperger's and IT: Dark Secret or an Open Secret*. www.computerworld.com.

McCarthy, Cormac. *No Country for Old Men*. New York: Vintage International, 2005.

Mumford, Lewis. *Technics and Civilization*. New York: Harcourt, Brace & Company, Inc., 1934.

Nielsen, Jakob. "F-shaped pattern for reading Web content." https://nngroup.com, 17 April 2006.

Pearcey, Nancy. *Total Truth*. Carol Stream: Crossway. 2008.

Postman, Neil. *Amusing Ourselves to Death*. New York: Viking Penguin, 1985.

—*Conscientious Objections*. New York: Vintage Books, 1988.

—*The Disappearance of Childhood*. New York: Vintage Books, 1994.

—*Technopoly*. New York: Vintage Books, 1993.

Ong, Walter. *Orality and Literacy*. London: Routledge, 1988.

Orwell, George. *1984*. New York: Signet Classics, 1961.

—"Politics and the English Language". *Horizon*. Vol. 13, issue 76, 1946.

O'Reilly, Charles et al. "Narcissistic CEOs and Executive Compensation". *The Leadership Quarterly*, 7 August 2013.

Ovid. *Metamorphoses*. Book VIII. 183-235.

Roberts James A., et al. "The invisible addiction: Cell-phone activities and addiction among male and female college students". *Journal of Behavioral Addictions,* 26 August 2014.

Rosengren, John, "Losing It All." *The Atlantic,* December 2016.

Shakya, Holly and Christakis, Nicholas. "A New More Rigorous Study Confirms: the More You Use Facebook, the Worse You Feel." Harvard Review, 10 April 2017.

Silberman, Steve. "The Geek Syndrome." *WIRED magazine* December, 2001.

Stein, Joel. "The Age of Trolls". *Time Magazine.* August 29, 2016.

Swingle, Mari K. *i-Minds.* Gabriola Island: New Society Publishers, 2016.

Twenge, Jean M. "Have Smartphones Destroyed a Generation?" *The Atlantic.* September, 2017.

—*iGen.* New York: Atria Books, 2017.

Vance, Ashlee. *Elon Musk: Tesla, Space X and the Quest for a Fantastic Future.* New York: Ecco, 2015.

Walton, John. *Genesis: NIV Application Commentary.* Grand Rapids: Zondervan, 2001.

Watts, Steven. *The People's Tycoon: Henry Ford and the American Century 2005.* New York: Vintage, 2006.

Weizenbaum, Joseph. *Computer Power and Human Reasoning.* San Francisco: W.H. Freeman & Co. 1976.

Wilder, Laura Ingalls. *On the Banks of Plum Creek.* New York: Harper & Sons, 1937.

CPSIA information can be obtained
at www.ICGtesting.com
Printed in the USA
FSHW020710040220
66773FS